CASE STUDIES IN AUTOMATION RELATED TO HUMANIZATION OF WORK

*Proceedings of the IFAC Workshop
Enschede, Netherlands, 31 October - 4 November 1977*

Edited by

J. E. RIJNSDORP

Twente University of Technology, Enschede, The Netherlands

Published for the

INTERNATIONAL FEDERATION OF AUTOMATIC CONTROL

by

PERGAMON PRESS

OXFORD · NEW YORK · TORONTO · SYDNEY · PARIS · FRANKFURT

U.K.	Pergamon Press Ltd., Headington Hill Hall, Oxford OX3 0BW, England
U.S.A.	Pergamon Press Inc., Maxwell House, Fairview Park, Elmsford, New York 10523, U.S.A.
CANADA	Pergamon of Canada, Suite 104, 150 Consumers Road, Willowdale, Ontario M2J 1P9, Canada
AUSTRALIA	Pergamon Press (Aust.) Pty. Ltd., P.O. Box 544, Potts Point, N.S.W. 2011, Australia
FRANCE	Pergamon Press SARL, 24 rue des Ecoles, 75240 Paris, Cedex 05, France
FEDERAL REPUBLIC OF GERMANY	Pergamon Press GmbH, 6242 Kronberg-Taunus, Pferdstrasse 1, Federal Republic of Germany

Copyright © 1979 IFAC

All Rights Reserved. No part of this publication may be reproduced, stored in a retrieval system or transmitted in any form or by any means: electronic, electrostatic, magnetic tape, mechanical, photocopying, recording or otherwise, without permission in writing from the copyright holders.

First edition 1979

British Library Cataloguing in Publication Data

IFAC Workshop on Case Studies in Automation Related to Humanization of work, *Enschede, 1977*
Case studies in automation related to humanization of work.
1. Man-machine systems - Congresses
2. Automation - Congresses
I. Title II. Rijnsdorp, J E III. International Federation of Automatic Control IV. Instituut van Ingenieurs. Division for Automatic Control
621.7'8   TA167   79-40431
ISBN 0-08-022012-6

These proceedings were reproduced by means of the photo-offset process using the manuscripts supplied by the authors of the different papers. The manuscripts have been typed using different typewriters and typefaces. The lay-out, figures and tables of some papers did not agree completely with the standard requirements; consequently the reproduction does not display complete uniformity. To ensure rapid publication this discrepancy could not be changed; nor could the English be checked completely. Therefore, the readers are asked to excuse any deficiencies of this publication which may be due to the above mentioned reasons.

The Editor

Printed in Great Britain by A. Wheaton & Co. Ltd., Exeter

IFAC WORKSHOP ON CASE STUDIES IN AUTOMATION RELATED TO HUMANIZATION OF WORK

Sponsored by
Committee on Social Effects of Automation,
International Federation of Automatic
Control

Organized by
Twente University of Technology,
Enschede, Netherlands

Contents

Summing - up the workshop
J. E. Rijnsdorp — ix

Program of the workshop — xi

1. General

Quality of working life: comments on recent publications in English
A.T.M. Wilson — 1

The design of work: new approaches and new needs
E. Mumford — 9

Work of the Social Effects of Automation Committee from Bad Boll to Enschede
R.M.J. Withers and J.E. Rijnsdorp — 19

2. Supervision of Processes and Systems

Human control tasks: a comparative study in different man-machine systems
C.L. Ekkers, C.K. Pasmooij, A.A.F. Brouwers and A.J. Janusch — 23

Basic transformations: the key points in the production process
C. Schumacher — 31

Jobs and VDU's, a model approach
S. Scholtens and A.J. Keja — 39

System development and human consequences in the steel industry
K.S. Bibby, G.N. Brander and T.H. Penniall — 45

3. Numerical Control, Assembly and Robots

Work organisation with multi-purpose assembly robots
M. Misul — 49

Man-Machine interface in the CONY-16 integrated manufacturing system
L. Nemes — 59

Automation and work organization - an interaction for humanization of work
J. Forslin — 67

Freeing the operator from the machine - as exemplified on assembly lines in the automotive industry
R. Knoll — 79

4. Other Subjects

Automation of WIG-welding
U. Lübbert — 87

Search of an appropriate transfer of technology to a developing region within an industrialized country
A. Dell'Oro, U. Pellegrini and C. Roveda — 91

An approach to the production line of automobiles by man-computer system
K. Namiki, H. Koga, S. Aida and N. Honda — 97

5. Late Papers

Two cases from the Norwegian process industry
A. B. Aune — 107

Distributed cellular manufacturing system
A. B. Aune — 115

Automation and humanization at the Dutch Railways; some examples
D. P. Rookmaaker — 119

Mechanization/automation and the development of the level of education and qualification of G.D.R. employees
H. Maier — 125

The socio-political responsibility of control engineers for the technical development of the future
E. Welfonder and K. Henning — 141

6. Discussion Report

Introductory Session — 155

Supervision of automated process — 157

Work organization — 161

Education and training — 164

Manufacturing and robot technology — 165

Automobile assembly — 167

Appropriate technology — 168

Summing-up and future — 169

7. List of Participants — 171

SUMMING-UP THE WORKSHOP

J. E. Rijnsdorp

Twente University of Technology, Enschede, Netherlands

Summary

The IFAC-Workshop on Case Studies in Automation, related to Humanization of Work was held at Twente University of Technology (Enschede, Netherlands) from October 31 to November 4, 1977. Experiences were discussed obtained in ongoing experiments in various countries, from which a clear picture was obtained of practical problems and constraints.
An important conclusion is that organization renewal is a necessary condition for humanization of work.
This volume of proceedings includes preprints, late papers, and discussion reports.

Introduction

As a sequel to the Workshop at Bad Boll (Schuh and Sprague, 1975), the IFAC Committee on Social Effects of Automation organized its Second Workshop at Enschede (Netherlands) from Monday October 31 to Friday November 4, 1977. The total number of participants was 62, from the following countries:

	Particip-ants	Papers
Netherlands	30	3
German Fed. Rep.	7	3
United Kingdom	6	5
Norway	5	2
Austria, Italy, Sweden, Switzerland (each 2)	8	3
Finland, France, German Dem. Rep., Hungary, Japan, U.S.A. (each 1)	6	3

About one third of the participants were social scientists, and two thirds control and automation engineers, hence there were good opportunities for exchange of points of view and experiences. In addition, positions and activities of labour unions were clearly presented.

The programme (see page III) contains 19 papers, amongst others in the field of supervision of automated processes (steel, plastics, chemical), mechanical and robot technology, automobile assembly, public transport, and appropriate technology.

New organization

In an introductory paper, prof. Enid Mumford indicated the assumptions, which are implicitly used by engineers in design activities. It appears that human factors score very low. She also stressed that every employee, also an unskilled one, shows:

- interest in work
- commitment to work
- desire and opportunity to increase skill
- desire and opportunity to participate
- work viewed as a meaningful part of life.

As became apparent from various discussions during the workshop, a necessary condition for satisfying these demands is a change from the traditional organization to a new one. This new organization can be characterized by:

- two-way communication
- co-operation between and integration of skills and monodisciplines
- flexibility in view of changing circumstances
- openness to the outside world

It has to fulfill the following functions:

- early introduction of all aspects and points of view in design projects, with special attention paid to worker participation and inputs from the unions
- autonomy for groups at all levels, for instance based on management by objectives.

These two functions can be related to democratization of work. The following ones are more akin to humanization of work:

- vertical job enrichment
- adaptation of technical degrees of freedom to "man" (the ergonomical approach).

Problem areas

Of course, it is not easy to renew existing organizations in the indicated way. A radical change is required of jobs and job attitudes at all levels of the organization. During the Workshop special attention was paid to the position of lower management, which can feel alienated and exploited in case of vertical job enrichment at the level below them. Another problem, which received less attention, is the compromise between open communication and efficiency.

A correct utilization of technical degrees of freedom is important in the design of worker-computer dialogues. This can be designated as humanization of interface software. The computer should be acceptable as a partner, instead of master or adversary. The computer can also be useful in vertical job enrichment, for instance as a tool for quality control, process initialization, optimization of operation, and scheduling. The dangers were indicated of extreme levels of automation. One has gone too far when the worker only has to wait for calamities, which require a high degree of alertness and skill at unexpected moments of time. It is questionable if an adequate human response still is possible under these circumstances.

In some automated processes workers tend to become isolated. One opinion is that introvert personalities might be happy under such circumstances. However, the other point of view, which was stressed during the Workshop, is that everybody has a need for social contacts, and work should provide possibilities for them.

Special attention should be paid to new technologies, which often are thrown into the market right after technical development. Then no or very little attention has been paid to human factors. This pertains, for instance, to process supervision systems with visual display units, and to industrial robots.

Economic limitations play a great role in the automobile industry, where liberation from the pace of the machine often requires expensive buffers between process steps. During the Workshop two films were shown: one about limited humanization of work in assembly operations at Daimler Benz (GFR), and one about the application of Volvo-Kalmar technology in an existing engine factory of Fiat (Italy).

Finally, attention was paid to macro-economical and macro-social effects. The contradiction between the increasing level of education and the de-skilling in many industrial jobs, the rapid obsolescence of skills, and the problems of reskilling, were shown in mutual relationships.

Conclusions

The Workshop has offered good opportunities to discuss progress in various ongoing experiments with humanization of work. The following main problems were indicated:

- organization renewal as a necessary condition for job design and ergonomical activities
- economic limitations
- providing sufficient and attractive work for people with a high level of general education

References

- P.A. Sprague and P. Schuh, Proceedings IFAC Workshop on Productivity and Man, RKW (Frankfurt, GFR. 1975)
- Newsletters of the International Council for the Quality of Working Life.
- Newsletters of the IFAC committee on Social Effects of Automation

Acknowledgment

The editor wishes to thank Mrs. Lena Mårtensson for her assistance and in preparing and contributions to the summing-up, and Mrs. Mimi Kroos and Miss Gea Middelbrink for the typing.

PROGRAM OF THE WORKSHOP

I <u>Introductory Session</u>

E. Mumford, The Design of Work, New Approaches and New Needs;
A.T.M. Wilson, The Quality of Working Life.

II <u>Supervision of Automated Processes</u>

C.L. Ekkers, C.K. Pasmooij, A.A.F. Brouwers, A.J. Janusch, Human Control Tasks: A Comparative Study in Different Man-Machine Systems;
K.S. Bibby, G.N. Brander, T.H. Penniall, System Development and Human Consequences in the Steel Industry;
A.B. Aune, Two Cases from the Norwegian Process Industry;
A.J. Keja, S. Scholtens, Jobs and VDU's; a Model Approach;

III <u>Work Organization</u>

C. Schumacher, Basic Transformations: The Key Points in the Production Process;
A.B. Aune, Distributed Cellular Manufacturing System;
D.P. Rookmaaker, Automation and Humanization at the Dutch Railways
J. Forslin, Automation and Work Organization: an Interaction for Humanization of Work;

IV <u>Education and Training</u>

H. Maier, Mechanization/Automation and the Development of the Level of Education and Qualification of GDR Employees

<u>Visit to N.V. Philips gloeilampenfabrieken, Eindhoven, Netherlands</u>

Presentation by H. Gelling about Humanization of work; visits to the studio for Electro-acoustics and to a Mechanical Workshop.

V <u>Manufacturing and Robot Technology</u>

R.M.J. Withers, J.E. Rijnsdorp, Work of the Social Effects of Automation Committee: from Bad Boll to Enschede;
M. Misul, Work Organization with Multi-Purpose Assembly Robots;
L. Nemes, Man-Machine Interfaces in the CONY-16 Integrated Manufacturing System;
U. Lübbert, Automation of WIG-welding;

VI <u>Automobile Assembly</u>

K. Namiki, H. Koga, S. Aida, N. Honda, An Approach to the Production of Automobiles by Man-Computer System;
R. Knoll, Freeing the Operator from the Machine as Exemplified on Assembly lines in the Automotive Industry;
H. Müller showed a film about the Fiat factory at Mirafiori (Italy)

VII <u>Appropriate Technology</u>

A. Dell'Oro, U. Pellegrini, C. Roveda, Search of an Appropriate Transfer of Technology to a Developing Region within an Industrialized Country

VIII <u>Summing-up and Future</u>

J.E. Rijnsdorp, Summing-up.
E. Welfonder, K. Henning, The Socio-Political Responsibility of Control Engineers for the Technical Development of the Future.

1. GENERAL

QUALITY OF WORKING LIFE: COMMENTS ON RECENT PUBLICATIONS IN ENGLISH

A.T.M. Wilson

London Graduate School of Business Studies, U.K.

Abstract. 'Quality of working life' and 'humanisation of work' refer to a growing cross-cultural and cross-disciplinary 'domain' of research, development, experimentation and discussion on the place of work in life as a whole, on the nature and relation of men to machines, and on the internal and external environment of work in its physiological, social, technological, economic and political aspects. These activities are inevitably multi-disciplinary and use concepts of many kinds. The paper offers examples, with brief comments, of eleven overlapping sources of recent and relevant publications in English: (i) from social philosophers and commentators; (ii) from economists and others, on the impact of technological developments; (iii) field studies by social anthropologists; (iv) ergonomics and allied disciplines; (v) sociologists and psychologists with interests in industry; (vi) engineers and technologists; (vii) research groups concerned with work-reorganisation and allied matters; (viii) analyses of actual and potential legislation; (ix) industrial executives and management associations; (x) labour union writers; (xi) policy statements by governmental and para-governmental agencies. There are concluding comments on situations and factors which foster or hinder diffusion of successful developments in this field.

Keywords. Economics; human factors; man-machine systems; optimisation; philosophical aspects; social and behavioural sciences; system theory. Not in the IFAC list are: quality of working life; humanisation of work; socio-technical systems; innovations in work organisation; industrial democracy.

INTRODUCTION

In Western European languages the various phrases used to describe this multi-disciplinary 'domain' represent not only conceptual and verbal differences; they also represent differences in regional, national and industrial cultures. In the Scandinavian countries, the notion of 'arbtsmiljö' - with its national variants - is used to cover the working environment in a particularly wide sense, including both social aspects of work and conditions of work. In practice, the apparently similar phrase 'Humanisierung des Arbeitslebens' does not refer to quite the same activities as the Spanish 'humanizacion del trabajo'. In French speaking countries 'amelioration des conditions de travail' is the most favoured phrase; in English-speaking areas 'improvement in the quality of working life' is again a little different, slightly more ambitious and perhaps less concrete. In these different descriptions there are, of course, wide areas of common concern, for in most countries the field we are considering invariably includes, for example, matters of safety, health and physical conditions at work (ILO, 1975; 1976; 1977); but there are also important differences in the boundaries given to the domain in any one country, or even by individual authors of papers - for example, over the intimacy or the separation between life at work and life in the household and the community; or in the varying significance given to large-scale socio-political differences between regions and countries; or to the different ways in which production of goods and services is related to social needs and culture patterns in different communities.

SOURCES AND TYPES OF CONTRIBUTIONS TO QWL STUDIES

The range of disciplines and professions concerned in this field means that an equally wide range of journals and publications are concerned. Integrative studies are usually the result of special occasions and often appear as reports of special conferences, e.g. of ILO, Geneva, OECD,

Paris, EEC, Brussels, or societies such as IFAC.

The list which follows gives eleven main 'types' and sources of publications in the language known as 'English'. These are briefly labelled and, with examples in each case, are intended to provide background by indicating a set of overlapping bounded areas from which, and within which, one can locate studies of QWL affairs of a more specific kind, particularly by writers with practical experience of developments in their field.

1. Publications by Social Philosophers and Commentators on Recent History

These frequently discuss 'alienation', a notion usually defined as "estrangement... between parts or the whole of the personality and significant aspects of the world of experience", e.g. estrangement of the worker from important aspects of the work situation. Publications on this topic obviously emerge from the continuous intellectual activity of academics and others, in their discussion and evaluation of human experience. Many of the more general aspects of such discussions are summarised, for example, in a special number of Current Sociology on "Alienation" - in essence a selected bibliography with a summarising and evaluative introduction by Peter Ludz (1970). Of a more specific character, an important discursive review (with bibliography) of "Alienation Studies" by Alvin Seeman (1975) begins with an illuminating statement on two contrasting orientations towards 'alienation' in those who write about it: "The enthusiasts make alienation the master concept - conveniently imprecise, empirically omnipresent, and morally irresistible when employed as a critique; while the doubters, with equal convenience, forget that dismissing the word in no way eliminates our dependence upon the 'root' ideas concerning personal control and comprehensible social structures which the alienation tradition embodies." In this 'review of reviews', Seeman briefly discusses the historical perspective and the varieties of alienation. He goes on to a descriptive and evaluative study of "Continuing Concerns" in the field. These include, for example, "The connection between the circumstances and the sentiments of alienation" which "...remains a substantial empirical problem - for example, the difference in alienation between the socialist states and Western capitalist societies..." He underlines the familiar difficulty that "alienation studies have remained entirely too dependent upon what is basically the "quick fix" - one-shot social survey results, with inadequate measures (one or two items is not rare), and with aspirations that do not match the realised gain..." "For those...made unhappy by this style of research...the possibilities in longitudinal research...have become more attractive...encouraged by the recent vigour of the 'social indicators' movement." This last point is exemplified by the work and publications in this field of the "Organisation for Economic Co-operation and Development" (OECD) in their "Social Indicator Development Programme", (1977).

2. Publications by Economists and Members of Related Professions on the Impact of Technological Developments.

In the 1960s there was an increased economic and financial interest in the relation of these developments to economic and social growth. The International Labour Organisation has a special branch on "Technology and Employment" and the Branch Chief, Ajit Bhalla (1976) has recently edited a series of papers on the topic.

An admired publication - it reached a second edition - was "Technical Change and Economic Development" by Salter (1969) which, in particular, drew attention to the complexity of these processes and the dangers of over-simplified approaches in studying them. A recent discussion of this last point is the OECD publication (1977) from a 'meeting of experts' in March 1976, on "Structural Determinants of Employment and Unemployment". The second of the four themes considered was "Capital-Labour Substitution, Technology and Employment"; and there is a summary of the six papers in this section, and of the discussion on them, by the general rapporteur on this theme.

At the level of the operating unit in industry and commerce two publications due this autumn give practical examples of the comparative costs of various forms of innovation in production units. Dr. Reinhold Weil of the Institute for Applied Work Sciences (IfaA), Köln, has material which expands on his recent monograph (1977) published by the International Institute for Labour Studies, on various aspects of new forms of work organisation; and Anthony Hopwood (1977) of the Oxford Centre for Management Studies, has completed a monograph in this same series: "Towards Assessing the Economic Costs and Benefits of New Forms of Work Organisation". There is now a marked trend to use 'value-added' concepts in evaluations of both labour productivity and capital productivity. A recent volume by a London-based consultant group (Urwick Orr, 1977) and published by the London "Financial Times" gives Western European comparisons on this basis. Two seminal papers in this area should be mentioned. The first, by Neal Herrick (1975) in "The Quality of Work and Its Outcomes: Estimating Potential Increases in Labour

Productivity"; the second, by B.A. Macy and P.H. Mirvis (1976), is "A Methodology for Assessment of Quality of Work Life and Organisational Effectiveness in Behavioural-Economic Terms". A highly significant point of a general kind, made some three years ago by Michael Shanks (1974) - then Director General of Social Affairs in the European Economic Commission - dealt with evaluation of economic results in new forms of socio-technical production unit. He said: "...a beginning might be made by taking a look at accounting conventions beyond those already forced on business and government by inflationary situations." The fact that accountancy is based on conventions has not always been remembered in the past; but he is here referring - for example - to the way in which costs of the results of stress at work have often been largely borne directly by the community and appear only indirectly in familiar organisational and company costing procedures.

3. Field Studies by Social Anthropologists

Initially describing older forms of society which differ greatly from industrialised communities, particularly over the nature and forms of work, these studies now use their approach, in part for comparative purposes, to 'contemporary societies'. The journal initially called Applied Anthropology and now called Human Organisation is a leading example of this literature. Related to this field but with a special interest in developing countries are the publications of the Intermediate Technology Group, whose journal is now called Appropriate Technology and whose founder, E.F. Schumacher, formerly economic adviser to the U.K. National Coal Board, has recently produced an influential paperback: "Small is Beautiful A Study of Economics as if People Really Mattered" (Schumacher, 1974).

4. Ergonomics and Allied Disciplines

A special issue of Ergonomics (Vol.19, No.3, 1976) - the journal of the international society of that name - contains a number of synoptic articles dealing with the broadening, and with past and future of these studies in different national communities. Much effort is currently devoted to the design of man-machine systems (e.g. papers in Human Factors; Ergonomics; Applied Ergonomics; IEEE Transactions, on Systems, Man and Cybernetics.) For bibliographic purposes this area can be linked to industrial medicine and to work concerned with safety and health which is reported in journals and regularly monitored and described by publications from sections of ILO,(1975, 1976, 1977) WHO, and national governments.

5. Sociologists and Psychologists with Interests in Industry

In 1972 an international conference sponsored by the Ford Foundation formally initiated the network and contacts between research groups, which is now expressed in the International Council for the Quality of Working Life. A revised set of the contributions to this conference has appeared in a two-volume work, "The Quality of Working Life" (1975), edited by Lou Davis, by professional origin an industrial engineer, and by Albert Cherns, a former Scientific Secretary of the United Kingdom Social Science Research Council. The first volume, sub-titled "Problems, Prospects, and the State of the Art", with twenty-eight papers, begins with an important "Assessment of the State of the Art" by the two editors. This discusses a wide spectrum of topics, from the attitudes of individual workers and professional groups to the changing orientations of government, employers and trade unions. There are three papers on technology and organisation which exemplify and stress that these are capable of reciprocal influence and joint optimisation. In one paper Professor Davis - if a short quotation from a wide context can be accepted with caution - expresses the criticism that "...engineers and technologists approach technical system and design as synonymous with or including organisation design. They see both as their proper and inherited domain of competence and responsibility... This inclusion of social system design is implicit, seldom if ever explicit, and is passed on to outsiders as a requirement of the technical system not to be questioned if the promised outcomes are to be realised." Conversely, in his own view, "...the design of technical systems has to be viewed as making choices among sets of opportunities available in the technology to permit the creation of jointly optimal socio-technical systems." The second volume, sub-titled "Cases and Commentary", has introductory chapters on problems of assessment of case studies in this field, followed by fifteen studies from both Europe and America.

Lisl Klein (1976) has published, in both English and German, a monograph study, invited by the Federal German Commission for Economic and Social Change, to provide a 'state of the art'report on the methodology of field studies of the design and redesign of jobs and the organisation of work in countries outside Germany. There was also an important commitment to give a critical commentary on developments in the field.

An extended series of books on quality of

working life is now being published by Martinus Nijhof of the Hague. The first volume, by Phil Herbst (1976) has the attractive title of "Alternatives to Hierarchies"; a later volume in the series, "Industrial Democracy: Form and Content", is a revised report and discussion by Emery and Thorsrud (1976) of their Norwegian programme of seminal research, in the late fifties and early sixties. The International Institute for Labour Studies (1976) closely associated with the International Labour Organisation, is now extending its series of monographs on innovations in work organisation in various European countries.

6. Engineers and Technologists

The publications of Professor Rosenbrock, Professor Rijnsdorp, and other members of the Committee who sponsor this IFAC workshop are likely to be known to its members; but it may be useful to underline the importance of the paper by Rosenbrock (1976) "The Future of Control", which clearly puts forward the major choices, and the results, of the different ways in which the inexpensive mini-computer can be used in design and in production. Under the present heading two further publications should be noted. The first, by John Burbidge (1975), is "A Study of the Effects of Group Production Methods on the Humanisation of Work: A Final Report", an extensive study for the International Labour Organisation, giving considerable detail on these developments. The second paper is by a technologist who is also a Fellow of the Royal Society of London - D.T.N. Williamson (1972). The Proceedings of that Society for 1972 carry his long, broad, constructive and critical analysis of contemporary forms of manufacturing, entitled "The Anachronistic Factory". His introductory summary contains these important points: "Industrial manufacture is the principle source of our national wealth, and it supports all other activities; consequently its wellbeing is of crucial importance to us all. It has been developing in three main streams - process manufacture, flowline or mass production, and batch manufacture - and of these only the first has kept in step with the parallel development of society. In the others, the trends of management organisation and the work patterns which this has created have been steadily veering into conflict with the values, aspirations and expectations of the people who have to make the industrial system work. The effects are visible now, and unless the patterns are changed, large sectors of industry could eventually become unworkable, because the qualities and skills needed in an individual are unlikely to be combined with a willingness to do the type of job offered." He reviews all forms of manufacturing system and, in particular, stresses the importance, in group technology, of the "fundamental principle...of pockets of self-contained responsibility in which man's skill, intelligence and enthusiasm are harnessed in somewhat specialised working groups."

7. Research Groups Concerned with Work-Reorganisation and Allied Matters

These often include faculty members of business schools as well as those of the special research centres and units recently set up in many countries. Their publications are often listed and summarised in special newsletters, which are greatly increasing in number through the rising costs and high volume of publications; and these, in turn, are beginning to exchange information about events, papers and publications - as do, for example, the newsletters of the IFAC Committee on Social Effects of Automation, and of the International Council for the Quality of Working Life. From business schools one would expect case studies; but these are only emerging slowly. In some cases the delaying factors are the complexity and delicacy of the negotiating situation between a company, management, its employees, their trade union officials and the external group seeking to report their developments.

8. Analyses of Actual and Potential Legislation

Many publications of this kind come from such organisations as ILO and EEC, or in the form of brochures or policy statements issued by governments to describe possible or actual programmes for legislation in this field. It is extremely important to recognise that the emphasis on regulatory activity by government, and its use in practice, varies greatly from country to country - for example, there is an accepted use, and a wide range of such legislation, in the Federal German Republic, with a marked contrast in the United Kingdom where, as Kahn-Freund (1975) stresses, the situation is very different. In the Scandinavian countries the political aspects of 'democratisation of work' remain prominent, with advanced programmes of training and retraining for those who will implement the new legislation and monitor its use. A recent phenomenon of importance - in France, for example - has been court decisions to imprison senior company executives who were held responsible for serious accidents at work.

9. **Industrial Executives and Management Associations**

Perhaps the most definite examples are those by Per Gyllenhammer (1977), Chairman of the Volvo company, and by the Swedish Employers Federation (1975). Such statements are, however, relatively cautious and infrequent - for example, through the market and economic uncertainties of recent years, or the difficulties of public discussion of the productivity potential of some development of work organisation or change of job content, in situations where national as well as company income and salary policies are matters of tense debate and negotiation. Many more experiments are under way in companies than have been publicly reported. In general, the 'professionalisation' of management is leading executives to develop an increasing concern with 'humanisation of work', if only to make it possible to attract and retain the kind of employees who will be able and willing to adapt to a turbulent environment. Conversely, difficulties of at least a temporary kind have arisen where the concept of 'self-managing groups' has been adopted as a policy element by a political party, as has occured in some countries.

10. **Labour Union Writers**

A recent circular from the United Kingdom Trades Union Congress (1976) to its member unions accepts the difficulties of organisational change at work in the current U.K. situation, but draws attention to the opportunities for modified socio-technical design where a new manufacturing unit is being set up. From trades unions one of the most outspoken positive statements on humanisation of work was recently made in the International Labour Review by a Vice-President of the United Autoworkers of America, Irving Bluestone (1977), who has negotiated a wide-ranging agreement for development in humanisation of work with General Motors.

11. **Policy Statements by Governmental and Para-governmental Agencies**

In recent years there have been outstandingly clear policy statements by the Dutch, Norwegian, Swedish, German and French governments. "Work in America" (1973), issued by the U.S. Federal Department of Health, Education and Welfare, reported an extensive enquiry and had a wide impact. A short brochure by Dr. N.A.B. Wilson (1973) "On the Quality of Working Life" preceded the setting up by the U.K. Department of Employment of a tripartite steering group and a research unit. Both the U.S. report and the U.S. brochure have valuable synoptic qualities.

FACTORS AND SITUATIONS WHICH FOSTER OR HINDER DIFFUSION OF SUCCESSFUL DEVELOPMENTS IN THE QWL FIELD

There are some familiar problems: 'goldfish bowl' situations of excessive and uninvited publicity for social experiments which ignores their need for some degree of 'protection' in their early phases; the problem of evaluation of results where there are 'Hawthorne effects' (i.e. the short-term response of operatives and executives to some general interest in their welfare, as against longer-term and more specific results of some technical or organisational change). There have been 'experiments' which fail - often through inadequate design - and are only reported by the 'grape-vine' of word-of-mouth communication where neither situations nor methods nor results are fully available for learning purposes. In relation to purely technical innovations - initially the use of new methods by U.S. farmers - Rogers (1962) outlined six stages in diffusion: awareness of some possibility; growth of interest in it; design of trial; consideration of results; acceptance or rejection. In QWL experimentation, where there is much greater complexity, wider considerations apply. (See Herbst, 1974) Analysis of recent experience (see Swedish Employers Federation, 1975) usually stresses that some methods of arousing interest (e.g. misleading claims for possible results of QWL experiments, or a high level of interest by groups outside the company) may lead to inhibition of the experiment in later stages through disappointment and resentment. There is good reason to believe that QWL experiments are most successful where they are developed, as it were, in the ordinary course of company activities concerned with development and adaptation, in which the main initiative, interest and involvement is of those directly and personally affected by the changes.

So far, most experiments have been in redesign of work organisation, with almost unchanged technologies; and the results show both similarities and differences in outcome in different communities and cultures; and there is need for comparative study, difficult as this can be. There are many lessons on this point from the results of projects organised by United Nations Agencies in different parts of the world. (Hyman, 1967)

Apart from data-processing and clerical activities, where there is a considerable literature on change processes and outcomes, by contrast there are, so far, very few full descriptions over time of experiments in manufacturing industry where a fundamental change in technology has been initiated at the same time as changes in work-organisation. Experiments of this kind have, however, been in progress in various countries for some time (for example, in coal mining, where the production situation and the newer technologies require multi-skilled teams with flexibility in their constituent jobs). However small in volume, however, the existence of such jointly-optimised socio-technical developments points to the need, and the possibility, of informing and interesting managing directors, design and production engineers, labour union officers and others in the concepts concerned. There is evidence of the increasing willingness of at least some members of these categories to consider the issues in this field.

It is equally obvious that there are considerable differences in attitudes towards these QWL experiments, in different countries, industries, companies, and categories of people in industry. In the latter two instances, the differences are often related to the 'philosophy' and 'culture' of the industry or the company; and there are equivalent contrasts and similarities in the orientation towards these developments of trades unions in neighbouring countries as well as between trades unions in the same country (Delamotte and Walker, 1974; Tchobanian, 1975; Rollier, 1977). Development of the capacity for 'organisational learning' is increasingly regarded as a prerequisite for success in QWL experiments; and this involves activities of various kinds in management education. The 'motivation to learn' will only develop if there is, throughout the company, a network of key people who are aware of its external situation, and of the possibilities and the necessities for change.

Finally, beyond the field of socio-technical restructuring at the level of the industrial enterprise there lie very large topics connected with the potential needs in most societies for redistribution of functions and tasks, as between almost traditional production systems, occupations, professions and disciplines; but on such matters needs and possibilities have not yet been clearly recognised.

REFERENCES

(The publications listed are only those mentioned earlier. Many of them contain bibliographies.)

1. Publications by social philosophers and commentators on recent history

Ludz, P. (1973). Alienation as a concept in the social sciences: a trend report and bibliography. Current Sociology, 21, 1, Mouton & Co., 7 rue Dupuytren, Paris 75006.

Seeman, A. (1975). Alienation Studies. American Sociology Review, 1, 91-123.

Organisation for Economic Co-operation and development (1977). Progress on social indicators. OECD Observer, March. 2 rue Andre-Pascal, 75775 Paris.

2. Publications by economists and members of related professions on the impact of technological developments

Bhalla, A.S. (1976). Technology and Employment in Industry: A Case Study Approach International Labour Office, Geneva. (English, Spanish). S.Fr.37.50. ISBN 92-2-101238-7.

Salter, W.E.G. (1969). Productivity and Technical Change. 2nd ed. Cambridge University Press, 200 Euston Rd., London N.W.1.

OECD Directorate for Social Affairs, Manpower and Education (1976). Capital labour substitution, technology and employment. In Structural Determinants of Employment and Unemployment. (Ta. 33831).

Weil, R. (1977). Alternative Forms of Work Organisation in Western Europe: Implications for Conditions of Work and Economic Efficiency. International Institute for Labour Studies, Case Postale 6, CH-1211 Geneva 22. (IILS/SAWO/7).

Hopwood, A.G. (1977). Towards Assessing the Economic Costs and Benefits of New Forms of Work Organisation. International Labour Office, Geneva. (In the press).

Urwick Orr & Partners (1977). International Manufacturing Performance and the Role of Technology. Financial Times Business Enterprises Division, 10 Bolt Court, Fleet St., London E.C.4. £50.70.

Herrick, N.Q. (1975). The Quality of Work and its Outcomes: Estimating Potential Increases in Labour Productivity. Academy for Contemporary Problems. Information about reprints from the author - 1501 Neal Ave., Columbus, Ohio, 43201, U.S.A.

Macy, B.A. and P.H. Mirvis (1976). A methodology for assessment of quality of work life and organisational effectiveness in behavioural-economic terms. Admin. Sci. Quarterly, June, 212-226.

Shanks, M. (1974). Closing summary, Conference on Work Organisation, Technical Development and Motivation of the Individual, Nov. 1974. (In ICQWL Newsletter Three, obtainable from Gen. Sec., ICQWL, London Graduate School of Business Studies, London N.W.1.).

3. **Field studies by social anthropologists**

Human Organisation, Q.J. of Soc. Appl. Anthropology, 1703 New Hampshire Ave., N.W., Washington, D.C. 20009.

Appropriate Technology, Q.J., Intermediate Technology Publications, 9 King St., London W.C.2.

Schumacher, E.F. (1974). Small is Beautiful: a study of economics as if people mattered. Abacus, London. Paperback.

4. **Ergonomics and allied disciplines**

Ergonomics (1976) 19, 3. Special issue on present and future of various branches of the subject. Taylor & Francis Ltd., 10-14 Macklin St., London W.C.2.

Human Factors. Bi-monthly. Johns Hopkins Press, Baltimore, U.S.A.

Appl. Ergonomics, Int. Q.J., IPC Science and Technology Press, 32 High St., Guildford, Surrey. GU1 3EW.

IEEE Trans. on Eng. Mg.ment, Q.J. of IEEE Inc., 345 East 47th St., New York, N.Y. 10017.

International Labour Office (1975). Occupational Health Problems of Young Workers. Occupational Safety and Health Series, No.26. (English, French). ISBN 92-2-101051-1.

International Labour Office (1976). Noise and Vibration in the Working Environment. Occupational Safety and Health Series, No.33. (English, French). ISBN 92-2-101526-2.

International Labour Office (1976). Accident Prevention. A Workers Education Manual. (English, French, Spanish). ISBN 92-2-101479-7.

International Labour Office (1977). The Protection of Workers against Noise and Vibration in the Working Environment. An ILO Code of Practice. (English, French, Spanish). ISBN 92-2-101709-5.

5. **Sociologists and psychologists with interests in industry**

Davis, L.E. and A.B. Cherns (1975). The Quality of Working Life. Vol.1: Problems, Prospects, and the State of the Art; Vol.2: Cases and Commentary. Free Press, New York, and Collier Macmillan, London. Available in paperback.

Klein, L. (1976). New Forms of Work Organisation. Cambridge University Press, 200 Euston Road, London N.W.1. Original version: Die Entwicklung neuer Formen der Arbeits Organisation: Internationale Erfahrungen und heutige Problemstellung. Otto Schwartz, Annastrasse 7, 3400 Gottingen, W. Germany.

International Series on the Quality of Working Life. Martinus Nijhof, Social Sciences Division, Leiden, The Netherlands.

    Herbst, P.G. (1976). An Introduction to Non-Hierarchical Organisations

    Emery, F.E. and E. Thorsrud (1976). Democracy at Work: Report on the Norwegian Industrial Democracy Programme.

    Bolweg, J.F. (1976). Job Design and Industrial Democracy: the Case of Norway.

    Emery, F.E. (1977). Futures We Are In.

International Institute for Labour Studies Research Series (1976). Geneva.

No.4 Weil, R. Alternative Forms of Work Organisation: Improvement of Labour Conditions and Productivity in Western Europe.

No.6 Delamotte, Y. The French Approach to Humanisation of Work.

No.8 Thorsrud, E. Democracy at Work

No.10 Burbidge, J. Group Production Methods and Humanisation of Work: the Evidence in Industrialised Countries.

No.11 Takezawa, S.I. Quality of Working Life: Trends in Japan.

No.14 Delamotte, Y. The Attitude of French and Italian Trades Unions to the Humanisation of Work.

No.15 Seashore, S.E. Assessing the Quality of Working Life: the U.S. Experience.

No.18 Rollier, M. Organisation of Work and Industrial Relations in Italian Engineering Industry.

(A second set of monographs on work organisation is being published by ILO, conditions of work department. They are in the press at the time of writing. Others in this series include Y. Delamotte, H. van Beinum, and A.T.M. Wilson.)

6. **Engineers and technologists**

Rosenbrock, H.H. (1970). The Future of Control. Address to 6th IFAC Conference, Boston, U.S.A. A shortened version of this paper appears in Automatica, 13, 4, 389-392. Pergamon Press, London.

Burbidge, J.L. (1976). Group Production Methods and Humanisation of Work: the Evidence in Industrialised Countries. IILS Research Series No.10. International Institute for Labour Studies, Geneva.

Williamson, D.T.N. (1972). The Anachronistic Factory. Proc. R. Soc., London. A.331, 139-160. Also in Personnel Review, 3, 3, Autumn 1974, London.

7. **Research groups concerned with work-reorganisation and allied matters**

IFAC Newsletter of Committee on the Social Effects of Automation. $\underline{1}$, H. Rosenbrock; $\underline{2}$, J. Dockstader; $\underline{3}$, B. Aune; $\underline{4}$, P. Schuh.

Other newsletters which exchange information include that of IfaA, Koln. They are listed in ICQWL Newsletter Ten, obtainable from Gen. Sec., London Graduate School of Business Studies, London N.W.1.

8. **Analyses of actual and potential Legislation**

Kahn-Freund, O. (1975). *Law and Labour - an Ambivalent Relationship*. Third Woodward Memorial Lecture, Department of Industrial Sociology, Imperial College of Science and Technology, 11 Prince's Gardens, London S.W.7.

9. **Industrial executives and management associations**

Gyllenhammar, P. (1977). How Volvo adapts work to people. *Harvard Business Review* July/Aug. 102-113.

Swedish Employers Federation (1975). *Job Reform in Sweden: Conclusions from 500 Shop Floor Projects*. SAF, Technical Dept., Södra, Blasieholmshammen 4A, Box 16 120, S-103 Stockholm 23.

10. **Labour union writers**

U.K. Trades Union Congress (1976). Circular letter to member unions, April 8th, on the importance of 'green field' sites for socio-technical developments. TUC, Great Russell St., London W.C.1.

Bluestone, I. (1977). Creating a new world of work. *International Labour Review*, $\underline{115}$, $\underline{1}$, Jan/Feb. International Labour Office, Geneva.

11. **Policy statements by governmental and para-governmental agencies**

W.E. Upjohn Institute for Employment Research (1973). *Work in America*. Report of a special task force to the Secretary of Health, Education and Welfare. MIT Press, Cambridge, Mass., U.S.A. ISBN 0-262-58023-3. Paperback.

Wilson, N.A.B. (1973). *On the Quality of Working Life*. A Report prepared for the Department of Employment. Manpower Papers No.7. Her Majesty's Stationery Office, London. SBN 11-360641-9.

FACTORS AND SITUATIONS WHICH FOSTER OR HINDER DIFFUSION OF SUCCESSFUL DEVELOPMENTS IN THE QWL FIELD

Rogers, E.M. (1962). *Diffusion of Innovations*. Collier Macmillan, New York and London.

Herbst, P.G. (1974). *Socio-Technical Design: Strategies in Multi-Disciplinary Research*. Tavistock Publications, London.

Hyman, H. (1967). *Inducing Social Change in Developing Communities: An International Survey of Expert Advice*. United Nations, Research Institute for Social Development.

Delamotte, Y. and K.W. Walker. (1974). *Humanisation of Work and Quality of Working Life: Trends and Problems*. Bulletin 11, International Institute of Labour Studies, Geneva.

Tchobanian, R. (1975). Trade Unions and the humanisation of work. *International Labour Review*, $\underline{3}$, $\underline{3}$, March. ILO, Geneva.

Rollier, M. (1977). *Approaches of Trade Unions Regarding New Forms of Work Organisation in Western Europe and North America*. International Institute of Labour Studies, Geneva. (IILS/SAWO/4). (Sig. Rollier is an Italian trade union official.)

THE DESIGN OF WORK: NEW APPROACHES AND NEW NEEDS

Enid Mumford

Manchester Business School, Booth Street West, Manchester M15 6PB

THE PROBLEM

Predictions about 'work' and what it will involve in the future abound. There are those who forecast that we are now entering the era of the 'post industrial' society in which a relatively small percentage of the labour force will be involved with production, assisted by a large amount of automated machinery, while the majority will be concerned with providing an ever increasing range of personal services associated with education, health, welfare etc. In Britain this forecast does not seem very credible. It is true that in certain industries, for example, chemicals, the white collar labour force is now larger than the blue, but we have also learnt to our cost that a service economy is an expensive one and requires an extremely profitable production sector to finance it.

An alternative forecast is that for a long time ahead we are going to require people to work in industry, designing products, making them and selling them, together with all the ancillary functions that these activities require, although the white collar/blue collar differences seem likely to become increasingly blurred. Therefore people, particularly young people must find industry an attractive place in which to work and believe that they will find interest, challenge and an opportunity to learn and develop their potential in the jobs that they are offered there. Unfortunately the evidence from most industrial countries at the present time is that this is not the case. Both on the shop floor and in the office people are beginning to rebel against the low level routinised work activities which they are required to carry out and which offer them little or no opportunity to use or develop their skills and talents. We appear to have been in the grip of a powerful ideology which has pervaded all industrial countries including both East and West in Europe, but which is now losing credibility. In Western Europe this ideology has seen labour as an expendable, easily replaceable commodity which produces at highest efficiency and lowest cost when few demands are made of it, when work is tightly controlled and when little or no discretion is allowed to the individual worker or to the work group. In Eastern Europe, either because engineers think in the same way or because some technology has been bought from the West, the physical organization of work appears similar and has resulted in the same tendencies towards work routinization.* Soviet commentators, like those of the West, believe that dull work has had adverse human consequences including an influence on workers to obtain work variety through changing their jobs. For example Dmitri Ukrainsky, Deputy Head of the department in charge of the introduction of new methods of planning and economic incentives of the State Planning Committee of the USSR Council of Ministers has said:

> 'In some parts of the country the drift of manpower in various branches of the economy reaches 25% of the overall workforce. This leads to major financial losses. The 25th CPSU Congress, which was held in February-March 1976, named labour fluctuation as one of the key problems that were to be resolved in the Tenth five year plan period, between 1976 and 1980.'(1)

But a new ideology is now appearing on the industrial scene and there are signs that many managers and trade unionists are accepting it. This is based on the belief that an alienated work force is an undesirable element in modern society and that this alienation can and should be reduced by considering the individual's work and work situation and ensuring that this is set up in such a way that he or she can do a job that is personally satisfying in a safe and comfortable environment. The reasons for

---

*This statement is derived from the Author's observations in Eastern European automobile plants.

the acceptance of this new ideology appear to be diverse. Some managers and trade unionists are subscribing to it because they believe that such a philosophy is morally correct, that people have a right to a good quality of life both inside and outside work and that in work this involves providing opportunities for personal development through enhancing knowledge and skill. Others may have less altruistic motives. Managers may perceive more interesting and attractive work as leading to higher levels of productivity or a reduction in industrial relations problems or a way of competing for scarce labour. Trade unionists may see a higher level of work responsibility as leading to higher earnings or more worker control at the operational level of the firm. However from the employees point of view it is the consequences of a humanistic philosophy for his own work situation that are important rather than the motives that lie behind the enthusiasm of either side in industry to subscribe to it.

THE DEVELOPMENT OF ALIENATING WORK

An excellent historical account of those ideas and problems which have led to the routinization and deskilling of work has been given by Williamson (2). He has described how in the latter part of the nineteenth and the early part of the twentieth century the needs of the Industrial Revolution resulted in the demand for skill becoming greater than the supply. As a result much engineering skill was applied to production problems which enabled manual skill to be diluted. F.W. Taylor separated the 'doing' of work from its 'planning', placing the latter firmly with management and removing variety and discretion from the former through techniques which gave a worker 'the one best way' to do a job. (3). New work study experts such as Bedeaux and the Gilbreths extended this approach producing analytical methods for splitting work into a number of basic and separable tasks which were then allotted to different workers. Williamson tells us.

> 'As a result of this pattern of thought, batch manufacture is almost universally accomplished today by issuing components into manufacture on an 'operation' basis, i.e. the work is split down by planning and work study methods into separate operations involving perhaps five to 30 operations per part. Each of these operations is treated separately, and may involve passing the part from one machine or process to another.....'

Williamson points out that this approach has not only reduced human job satisfaction by creating routine and boring work it has reduced efficiency by producing queuing problems and idle component time. But not only do unskilled people have to do these dull jobs, this is also true of the skilled worker who may shape pieces of metal without knowing how his contribution fits into the final product and without the opportunity for showing originality or improving work methods. 'He works within a rigid structure in which someone else has determined what he should do and how he should do it'.

Both the flow line principle and the assembly line based on machine paced work have added to the human problems. Batch work based on single purpose machines could produce routine but it left the worker in control of his machine and able to start and stop it. The moving assembly line adds pressure to monotony with the machine now in command and determining the speed at which the human being works. The author has worked on assembly lines where, whenever management wished to raise production, the speed of the line was arbitrarily increased without the workers being given any warning. Because of this the workers were placed under considerable stress although, needless to say, they had their own methods for retaliating against management.

Process technology brings its own human problems with workers frequently isolated from one another on a large plant or required to do very little for long periods of time and then take urgent and skilful action when a breakdown occurs.

Many of the ideas that originated with shop floor production systems are now being applied in the office. The breaking of work up into small simple tasks is general and the flow line principle is common although happily without the intervention of machine pacing. Office technology is bringing its own problems. Computer systems which are poorly designed in human terms can introduce new routine jobs particularly at the data input stage, while the clerk's ability to undertake a logical sequence of clerical tasks can be interrupted through the intervention of a computer half way through the work process. Computers also provide management with opportunities for introducing tight control systems that monitor work output. The worker on the shop floor may have his machine linked to a computer which notes when it is shut off. The office worker may have his errors and his speed at processing documents identified and listed. The key punch or VDU operator may have her number of key depressions per hour automatically recorded.

The cause of this deterioration in the human work environment is largely, but not entirely, due to the acceptance by management of a set of powerful ideas on how work could best be organized to reduce costs and increase output. However it must be recognised that trade unions have also contributed to this erosion of work function as a result of their efforts to protect the interests of their members through demarcation and the restriction of certain types of work to certain categories of worker.

Design Principles

The design principles commonly used by management were investigated by Davis, Cantor and Hoffman in the 1950s, who found that companies tried to.

1. Break the job into the smallest components possible to reduce skill requirements.

2. Make the content of the job as repetitive as possible.

3. Minimize internal transportation and handling time.

4. Provide suitable working conditions.

5. Obtain greater specialization.

6. Stabilize production and reduce job shifts to a minimum.

7. Have engineering departments, whenever possible, take an active part in assigning tasks and jobs. (4)

Davis concluded that for the engineer the usual indicator of achievement is minimum unit operation time. He obtains this through specifying the content of work so that.

1. There is specialization of skills.

2. Skill requirements are minimized.

3. Learning time is minimized.

Individual tasks are combined into specific jobs so that.

1. Specialization of work is achieved whenever possible by limiting the number of tasks in a job and limiting the variations in tasks or jobs.

2. The content of work is as repetitive as possible.

3. Training time is minimized.

A follow up study by Taylor in 1976 suggests that these principles still have a great deal of support from engineers. (5)

ATTEMPTS AT AMELIORATION

Work Reorganization Around Technology

To date attempts to improve the quality of the work situation for the workers have taken a number of forms, almost all of which have focussed on different ways of organizing and allocating task structures around a particular technology which is regarded as immutable. On the shop floor, improvement of the work situation has generally been attempted through enriching jobs by adding to a set of production tasks introductory activities such as material requisition and tool setting, final activities such as inspection and servicing activities such as machine maintenance. Effectively this has amounted to returning to the worker what he or she used to have before these ancillary activities were removed as a result of a managerial belief that the smaller the job the faster and more efficiently the worker would work and the less the firm would have to pay for labour. In some countries, including Britain, although this approach could improve the quality of the work situation it may be difficult to introduce because of the different interests of shop floor unions. For example the British part of an international study of automation in the automobile industry has recently examined the differences in the work content of machine operators and ancillary staff such as supervisors, maintenance workers, setters and section leaders on jobs that have been defined as of low automation, medium automation and high automation. (6) The low automation jobs covered single and multi-purpose machines used for cutting, grinding and turning operations, together with some routine jobs associated with controlling transfer lines. The medium automation jobs were associated with early transfer line technology which required considerable manual intervention; the high automation jobs were located on more advanced transfer lines. The results showed only marginal differences between the machine operating jobs at each automation level in terms of such factors as variety, independence, responsibility, use of knowledge and training, creativity, learning, work interest and opportunity to develop abilities. The differences between the supervision/maintenance and setter/section leader groups and the machine operators were much greater, irrespective of the level of automation. The jobs of the machine operators both on single machines and on the transfer lines could clearly have been enhanced by giving them maintenance and setting responsibilities, together with some of the tasks handled by section leaders and supervision. But it seems unlikely that the unions involved would have accepted such reorganisation.

Another approach to enriching shop floor work was pioneered in the British coal industry in the fifties and is based on the formation of autonomous groups (7). Each of these groups is multi-skilled and takes responsibility for a substantial piece of production activity. In the coal mining example it was for all the activities associated with getting coal from a particular section of coal face; activities which were normally carried out by three different groups of miner's, carrying out different tasks on different shifts. The autonomous, or as it is now sometimes called, socio-technical or self managing concept required each group to assume responsibility for the management of this piece of face. This involved the allocation of tasks and shifts on the basis of mutual agreement together with a joint problem solving approach by the team as a whole when difficulties were encountered. Interaction with management was largely restricted to arriving at an agreement on how much coal the work group would produce and how much they would be paid for doing this. Since this first experiment the approach has been tried out in a number of other countries and industries. The chemical firm Norsk Hydro was a pioneer in Scandinavia and Dmitri Ukrainsky reports on experiments in Russia following economic reforms ten years ago (8). He says.

> 'The idea of semi-autonomous work teams came into being and gained support. Such teams are called contract teams in building projects and work-order-free squads in agriculture. In the engineering and chemical industries they have no name coined for them to date but they have accumulated a collective experience in handling a rather important socio-economic problem.
>
> These work groups are usually paid upon completion of their job by results (the payment is made to the whole team) and have considerable independence granted to them in the organization of the job and its apportionment among the members. They find the "fatal" division into machine-operators and maintenance men unsuitable since the overall wages directly depend on the productivity and skill of each worker. They are interested in having each new team member quickly develop trade skills. But the skilled man does not think it beneath his dignity now and then to perform subsidiary operations if this benefits the job.'

Some attempts have also been made to improve the quality of working life for workers on assembly lines, a very large proportion of whom are women, through reorganizing work. Many years ago experiments were carried out in French industry which permitted women working on these lines to control the speed at which the line was running (9). It was recognised that the natural speed at which people worked varied throughout the day and also varied according to physical factors such as temperature, yet the logic of the moving assembly line was that all workers could work at the same unvarying speed from morning to night. The Philips plant at Eindhoven later introduced the concept of 'buffer stock' onto their lines so that each girl, after completing her own assembly operation slid the component onto a table beside her from which the next girl took it. This 'buffer' enabled girls sitting next to each other to work at different speeds. Philips later extended its experiments by increasing the size of the assembly task so that instead of assembling a small part of a television set each girl was able to assemble a major part. Other attempts at ameliorating the assembly line problem have been through giving the work group responsibility for planning and inspection activities that might normally be carried out by another group or by supervision. However all these efforts are palliative and do not tackle the basic problem of this kind of work which is that the worker is paced by the machine and controlled by the machine, rather than vice versa. Few firms yet appear to have challenged the basic concept of assembly line technology. For example, by replacing the moving production line with a material supply system that helps the worker yet enables him to work at his own speed on a comprehensive piece of assembly. The German trade union I.G. Metall has recently contributed to such a change by negotiating a minimum task cycle of 90 seconds. This is a much longer task cycle than is found on most female assembly jobs.

In the office we find exactly the same kind of approach to improving the quality of working life. Despite the impact on the office scene of a new form of technology - the computer - a technology which appears to be less deterministic than most other technologies, we find that design of the technical part of the office system has still tended to be based on the traditional engineering concept of optimisation of machine potential. A considerable number of experiments in improving the quality of working life have taken place in the office situation, and have ranged from individual job enrichment, usually through permitting the individual employee to take over a meaningful portion of the service process and to look after many of the needs of a group of customers, to the autonomous or self-management group concept in which a group of employees are given responsibility for a major sector of work and encouraged to acquire multiple skills, to rotate work, to take responsibility for product or service quality and for the planning of work. However here, as on the shop floor, the major changes in the human organization of work have taken place as a rule without any comparable major changes in office technology.

Work Reorganization in Parallel with Technological Reorganization.

Group technology. Although most attempts to improve the quality of working life have regarded the machine part of the system as untouchable and not to be questioned, there have been some attempts to associate social reorganization with some alteration of the technology or vice versa. Perhaps the best known and most successful of these has been the method of organizing production work called group technology or the 'cell' system which also lends itself to a form of autonomous group organization. Group technology was a Russian invention pioneered by S.P. Mitrofanov primarily as a method for improving machine tool utilization by classifying and grouping 'technologically similar' groups of parts together for machining (10). A number of firms, in particular Serck-Audco Valves in Britain took the approach a logical step further and set up separate manufacturing cells based on machine 'families'. Any machine family 'cell' dedicated to the production of related components - that is a cell containing the necessary drilling, milling, turning and other machines required to produce the components - requires a group of workers with a variety of skills. It is logical for such groups to become multi-skilled and to assume responsibility for their own planning, error correction and inspection. Some firms give their cells the resources they need for a month's production and then leave them alone (11). The potential of such 'cells' to provide responsibility, challenge and work interest should therefore be very great. Unhappily, research carried out for the Mechanical Engineering NEDO in 1974 showed that these opportunities were rarely grasped (12). Most firms introduced group technology to secure technical efficiency gains and the opportunity for making human improvements was largely let pass. There has been little attempt to improve participation, to devolve decision making to cell groups or to promote autonomous groups in which are small, self co-ordinating and in which the members have multiple skills. The author was recently shown the group technology section of a firm and found that this differed from the normal flow line system only in the sense that there was more than one type of machine. Workers were carrying out small, routine jobs without job rotation or any additional work responsibilities.

Group technology does not require any new concepts of machine design, although it does incorporate some new concepts of machine location as it replaces the flow line concept with a cell arrangement in which groups of workers are associated with 'families' of machines. The concept of the autonomous or self managing group is therefore very easy to apply to group technology. A more advanced version of this concept might be to replace the machine family with multi-purpose machines on which the operator is able to carry out all the required machining operations. The use of such a machine would not only require machining skill, it would also require setting and maintenance skills of a high order. A problem here is that in the future complex machines are likely to be increasingly numerically controlled and the trend to date has been to transfer the use of skill from the operator to the programmer who writes the program for the machine (13). Our new concept of work requires that the operator should be responsible for and competent to carry out his own programming.

It must be recognised, of course, that there is often no demand from the workers themselves for increased interest or responsibility. In the British contribution to the international study of 'Workers and Automation' the automobile workers gave an improvement in working conditions and safety as their most urgent requirement. It can be argued that in many countries, including Britain, greater work interest is not stated as a need, first, because workers do not see any possibility of their production systems being reorganised to provide this; second, because their trade unions do not bring quality of working life, other than working conditions, into the bargaining area; third, because work interest can often be achieved through fighting the management; fourth, because when work is instrumental in the sense that it provides little intrinsic satisfaction other than pay, it is viewed as instrumental and job satisfaction is not seen as a realistic objective.

Computer technology. Computer technology fits into the category of technology which is in its nature flexible enough to be adjusted to meet human needs. Today, its human effects seem to a large extent, although not entirely, to be due to systems design decisions that are taken by the firm introducing the computer application, rather than to any inherent qualities of its machine base (14). With computers it is possible to take decisions on what should be automated and what left alone. In other words on where to place the man/machine boundary. An approach that would fit the philosophy outlined here is that automation should be used to handle work that is routine and undemanding and cannot be enriched in any way. It should also be used for work that is dangerous or physically or mentally damaging. Attempts should not therefore be made to get computers to take complex decisions, or to carry out complex operations, these should be given to the human being. Unfortunately the technical designers' present set of values, together with their liking for working at the frontiers of knowledge have often made them neglect the automation of routine work and concentrate instead on automating aspects of work that are of a higher level. It can be suggested that the likely shift from large centralised computer systems to distributed systems, irrespective of whether these are based on small machines or interactive, on line, terminals, provides us with a new opportunity for formulating human values in respect of computers. For example, the autonomous group concept can be assisted through the computer's ability to provide work groups with accurate, up to date information which they can use to monitor their own performance and set their own targets. Computer technology, unlike most other technologies is extremely flexible and its consequences are far more a product of design decisions on how to <u>use</u> the technology than decisions taken at the initial design stage. Exceptions to this generalisation are to be found with data input where machine design imposes a rigid task structure on key punch operators and replicates the anti-human environment of the worst kind of shop floor work. Data input requires large groups of girls to work on very routine kinds of machine jobs punching data onto paper or magnetic tape. Even the intervention of the VDU screen with the possibility of reading and checking what is recorded, does little to increase work interest, unless the screen has an enquiry facility. One of the Norwegian unions is now attempting to reach agreement with Norwegian management that VDU screens shall be two way. That is the operator must be able to ask questions of the computer and obtain information from it and not be restricted to inputting information only.

Technological Design which Facilitates Work Re-organization

A fundamental means for reducing the risk of alienated workers is to ask engineers to accept responsibility for the human consequences of their design decisions. That is to ask them to recognise at the prototype stage of machine design that technically sophisticated machines may mean bored and frustrated workers and that ideally skill should be shared between machine and man. Some years ago a British sociologist, Joan Woodward, suggested that engineers ought to accept responsibility for the industrial relations problems that arose as a consequence of their shop floor decisions. May we ask them to do the same with their design decisions?

The most exciting contributions to the flexible design of technology so that new forms of work organization can be assisted rather than inhibited have come from the Scandinavian firms Volvo and Saab Scania. Pehr G. Gyllenhammar, president of Volvo has said.

> 'At Kalmar, the objective is to organize automobile production in such a way that employees can find meaning and satisfaction in their work.
>
> This is a factory that, without any sacrifice of efficiency or financial results, will give employees the opportunity to work in groups, to communicate freely, to shift among work assignments, to vary their pace, to identify themselves with the product, to be conscious of responsibility for quality, and to influence their own work environment.
>
> When a product is manufactured by workers who find their work meaningful, it will inevitably be a product of high quality' (15).

Technical flexibility has been built into the production system through the development of battery powered assembly carriers that function both as transport devices and assembly platforms. These move on magnetic tracks embedded in the floor but can be guided manually with operating levers. This assembly carrier makes it possible to choose between various organizational forms in the planning of assembly work. In other words it caters for a choice of work organization so that either an autonomous group structure or a more conventional straight line assembly approach is possible with the same technology. In addition VDUs provide work teams with information on quality variations so that work methods can be altered by the team itself if quality is not meeting required standards.

Another major contribution to thinking on technical design has come from the shop stewards of the British aerospace company, Lucas (16). In a corporate plan prepared by the shop steward committee they provide another 'quality of working life' argument by suggesting that not only do workers want challenge and interest, they also want to work on socially useful products. The shop stewards state.

> 'The desire to work on socially useful products is one which is now widespread through large sectors of industry. The aerospace industry is a particularly glaring example of the gap which exists between that which technology could provide, and that which it does provide to meet the wide

range of human problems we see about us. There is something seriously wrong about a society which can provide a level of technology to design and build Concorde but cannot provide enough simple urban heating systems to protect the old age pensioners who are dying each winter of hypothermia'.

The shop stewards committee also have some strong comments about the design of technology and deplore attempts to replace human intelligence by machine intelligence. Their suggestion for overcoming the problem of a work situation that is deteriorating in human terms is to bring workers into the design and development stages of new products. They recommend the setting up of integrated product teams which incorporate design, development production engineering and manufacturing. They suggest that this approach will result both in the development of socially useful products, and the protection and enhancement of work and of the work environment. Such an approach would involve considerable retraining and the shop stewards point out that here they are not concerned only with the retraining of white collar and technical staff. They believe that the entire work force including semi-skilled and skilled workers are capable of retraining for jobs which would greatly extend the range of work they could undertake.

So far the design and use of technology has been discussed as an influence on the quality of the working environment. But technology is also of major importance in those parts of the developing world where major unemployment is a continuing problem and the issue there is one of substituting labour for capital, together with making appropriate technical choices. This too has its design implications. The need of the developing world may be less for intermediate technology than for technology which is sophisticated, yet simple to use and which enhances the skill of the worker but does not replace it (17).

THE SOLUTION - A CHANGE IN VALUES

Technology is unlikely to be changed unless groups with some power and influence recognise and accept a need for it to be changed. This applies to management, trade unions and to the workers themselves. There are still very few strikes over the routine nature of work and few trade unions are yet making work interest and challenge one of their bargaining issues. There are however some exceptions to this last comment. The Norwegian National Union of Iron and Metal Workers (MWU) is providing educational courses for its members to enable them to present an informed point of view when company policy decisions on the use of technology are being taken (18). In Britain, however, it is still rare for the trade union officials to participate in any meaningful way in the design of technology, other than showing concern over redundancy, grading and salary issues.

A change in values requires some statement of new values and here the philosophy of John Dewey in his book Reconstruction in Philosophy written in 1949 may be useful. Dewey states:

> 'Government, business, art, religion, all social institutions have a meaning and purpose. That purpose is to set free and to develop the capacity of human individuals without respect to race, sex, class or economic status. And this is all one with saying that the test of their value is the extent to which they educate every individual into the full status of his possibility'(19).

It may also be useful to try and translate this statement of values into a set of principles for the design of technology. As a start one might suggest that any technical system should provide individual workers with opportunities for skill, problem solving, control, personal development and social relationships. It should also be flexible so that different forms of work organization can fit easily with it.

The design of machines or machine systems should, if possible, assist the following. It should certainly not prevent them taking place:

1. The development of a skill which provides the worker with job interest, a sense of challenge, a feeling of competence and a desire to learn more.

2. Following on from 1 above. The development of a progression of skills ranging from comparatively simple to complex. If possible the technology should assist the operator to train himself so that he can learn as a result of his own interaction with the machine.

3. The operator's own control over the machine and the machine environment. The operator should be able to start, stop, control the speed of the machine and make necessary adjustments to it. Machines which control the operator because they are pre-programmed in some way should be regarded as ethically unacceptable.

4. The operator working as a member

of a small, supportive social group. Isolated, individual tasks, such as, for example a haulage worker in a coal mine with responsibility for keeping a section of an underground conveyor belt working, should be regarded as unacceptable. Similarly, machines which require routine work from the operator but also concentration should be regarded as unacceptable for social reasons as well as on the grounds of 1 and 2 above.

5. Machines and machine systems should be flexible enough to be able to be adjusted if the operator develops new methods and techniques which he or she considers to be more effective in the sense that they produce a higher quality product or help avoid the occurrence of work problems. This kind of improvement should be recognised to be a responsibility of the individual worker or the work group and not solely that of a work study or O & M department.

The points listed above are based on a set of long term values related to enabling a worker to use and develop his potential. Other values may be a product of particular economic problems that arise from time to time. For example, Britain at the moment has a great need to solve its unemployment problems, particularly those of young people trying to enter work for the first time. Two further principles might assist these unemployment problems.

6. Automation should not be encouraged as an end in itself. Consideration should be given to forms of technology which are not labour saving, providing they meet the skill and learning criteria listed earlier.

7. Principle 2 above could be used as a method for assisting the training of young workers without the constraints of expensive and time consuming apprentice schemes. This is not to suggest that apprentice schemes should be done away with but that there should be alternative methods for acquiring skills which are open to the young workers who are unable to be accepted as apprentices.

CONCLUSION

This paper has posed a number of serious problems concerning the relationship between man and work and it has examined existing solutions and future needs. Although research into the design of work has always taken place with new solutions emerging to solve new problems there are research topics which could fruitfully be pursued at present. Some of these are listed below.

1. More knowledge of what different groups are seeking from work in terms of job satisfaction. Does every worker want interest and challenge or do some prefer routine and structure?

2. An understanding of the extent to which existing technology has a potential for flexibility so that different forms of work organization can be fitted to it. Is an autonomous group compatible with an engine block transfer line, for example?

3. An understanding of the ability of automation, both in the shop floor and in the office, to assist flexible work organization. Can automation be a liberator and not a tyrant?

4. An understanding of the relationship between design decisions taken when new machinery is being evolved and the quality of working life of the workers who will eventually use these machines.

However research will not contribute to change unless there is a wish to change and here society's values will be the dominant influence. The implementation of new values in turn requires new information so that future problems can be identified and averted. Democratic methods for technological assessment can make a contribution here.

Finally it is appropriate to end with a comment from Professor Louis Davis, an engineer and a social scientist who has had a major influence on the design of work. He says.

> 'Engineers must recognise that in designing technical systems, they are also designing social systems, and that the assumptions about man which underlie their designs are often horrific' (20).

REFERENCES

(1) D. Ukrainsky, To each according to his interests, *Sputnik* (Digest of the Soviet Press), April 1977.

(2) D. T. N. Williamson, The anachronistic factory, *Proceedings of the Royal Society*, 1972. Reprinted in *Personnel Review*, 4, 26 (1973).

(3) Taylor, F.W. (1911) *The Principles of Scientific Management*, Harper, New York.

(4) L. E. Davis, R. R. Cantor, J. Hoffman, Current job design criteria, *Journal of Industrial Engineering* 6, 5 (1955).

(5) J. C. Taylor, *Unpublished* pilot survey (1976).

(6) P. Lloyd, S. Mills, Automation and industrial workers, Final Report to Social Science Research Council, *Unpublished* (1977).

(7) Trist, E. L., Higgin, G. W., Murray, H., Pollock, A. B. (1963) *Organizational Choice*, Tavistock Publications, London.

(8) D. Ukrainsky, ibid.

(9) Friedmann, G. (1946) *Problaimes Humains des Machinisme Industriel*, Gallimard, Paris.

(10) Mitrofanov, S. P. (1966) *Scientific Principles of Group Technology*, National Lending Library, Boston Spa, Yorkshire.

(11) Edwards, G. A. B. (1966) *Readings in Group Technology.* The Machinery Publishing Company, London (1971).

(12) S. Mills, The social consequences of using group technology in batch machining work shops in the mechanical engineering industry. *Unpublished* report for Mechanical Engineering NEDO, London (1974).

(13) Christensen, E. (1968) *Automation and the Workers*, Labour Research Department, London.

(14) E. Mumford, Job satisfaction, a major objective for the system design process, *Management Informatics*, 2 (1973).

(15) Aguren,S. Hansson,R. Karlsson,K.G. (1976) *The Volvo Kalmar plant.* The Rationalization Council SAF-LO, Stockholm (1976).

(16) Lucas Aerospace Combine Shop Steward Committee, Corporate plan : a contingency strategy as a positive alternative to recession and redundancies, *Unpublished* (1976).

(17) Bhalla, A. S. (Ed) (1975) *Technology and Employment in Industry,* I.L.O. Geneva.

(18) K. Nygaard, O. T. Bergo, The trade unions - new users of research, *Personnel Review* 4, 5 (1975).

(19) This is quoted by Sackman, H. (1967) in *Computers, Systems Science and Evolving Society,* Wiley, New York.

(20) L. E. Davis, The coming crisis for production management: technology and organization, *International Journal of Production Research,* 9, 65 (1971).

WORK OF THE SOCIAL EFFECTS OF AUTOMATION COMMITTEE FROM BAD BOLL TO ENSCHEDE

R.M.J. Withers* and J.E. Rijnsdorp**

Urwick Technology Management, London, England
**Twente University, Enschede, The Netherlands

Abstract. The paper summarises the activities of the IFAC Committee on Social Effects of Automation during the period 1974-1977.
— A Workshop was held about "Productivity and Man" at Bad Boll (FRG) in January 1974;
— A number of Newsletters were published;
— Factory visits were made to a Steel Rolling Mill, an Automobile Plant, an Electro-Mechanical and an Electronic Assembly plant;
— Contributions were made to the IFAC Congress at Boston (August 1975), and will be offered to the Congress at Helsinki (June 1978);
— Preparations are being made for a Workshop on Case Studies in Automation, related to Humanization of Work, to be held at Enschede Netherlands in November 1977.

Keywords. Social and behavioural science; Man-machine systems; Robots; Manufacturing Processes; Job design; Group technology.

INTRODUCTION

The International Federation of Automatic Control was founded during an International Conference at Heidelberg (1956). It is supported by National Member Organizations from all continents.

Every three years a general congress is held', where all fields and aspects of control are included. The interpretation of the term "control" has gradually become broader; it now encompasses the interactions with planning, operations research, world dynamics, the environment, economics, computer science, and human factors.

IFAC also has a number of technical committees, which are responsible for planning and coordinating activities in certain fields, such as control theory, education, applications, manufacturing technology, systems engineering, management science. In 1971 a Committee was formed to deal with Social Effects of Automation. The decision was made to direct the attention primarily on a specific area: the interactions between automation techniques and conditions in industry.

More specifically the aims were to make all control engineers aware of social effects of their work, to involve a smaller group of interested people in a more thorough analysis of social effects, and to provide information about new developments

WORKSHOP ABOUT "PRODUCTIVITY AND MAN"

The first major event sponsored by the committee was the Workshop in Bad Boll (FRG) in January 1974. The proposal for such a Workshop was made at the Paris conference of IFAC in June 1972, during a round table discussion.

The workshop has been fully described by Peter Schuh and Phil Sprague in an official publication "Productivity and Man". It had the following main characteristics:

— provision of ideal facilities in the Evangelische Acadamie at Bad Boll

— attendance by forty-three men and women from twelve countries

— representatives from trades unions, behavioural science, engineering and management

— presentation and discussion of nineteen pre-printed case studies in four sequential work sessions

— an introductory session to set the main topics and a closing session to produce a concluding statement

— publication of the Workshop proceedings

The Concluding Statement of the Workshop reads as follows:

"Based on (a) the case studies submitted from thirteen countries documenting the application of automation and technology to a variety of industrial and institutional problems and (b) the three days of discussions by our group of more than forty participants representing a wide range of educational and professional disciplines, we make the following observations:

1. The case studies submitted indicate that great gains in productivity have been achieved through the application of automation to human activities; these gains have not only been realized in terms of greater output, increased wages and improved product quality but also in extended leisure time, improved health and better living conditions.

2. The case studies submitted reveal that in many cases too little consideration was given to human factors of work in the design of the automation systems.

3. The Workshop participants concluded that the human shortcomings mentioned in point 2 above were a result of (a) insufficient recognition of the opportunities inherent in new technology to overcome these ill effects and (b) inadequate recognition of the increased awareness of people regarding the dehumanization of work which can result from the application of automation.

4. The Workshop participants believe that man/machine relationships can and should be optimized and that, therein, lies the greatest opportunity and challenge for the control engineer.

5. The Workshop participants believe that, too frequently, the control engineer is brought into the design planning process too late to make an optimum contribution; further, that there has been insufficient dialogue and co-operation between control engineers and social scientists due primarily to their respective ignorance of each other's discipline and education.

Recognizing our primary responsibility to the International Federation of Automatic Control and the control engineers and scientists throughout the world which make up its membership, we make the following Recommendations to Control Engineers:

1. In designing automation systems, the Control Engineer should consult with and encourage the active participation of all people who are or will be involved in the system.

2. In designing automation systems, the Control Engineer should not restrict the amount of information about the system; on the contrary, he should provide all people involved in the system with as much information as possible.

3. In designing automation systems, the Control Engineer should consult and co-operate with suitably qualified social scientists and trade union representatives in order to produce more effective systems from a human standpoint.

4. In designing automation systems, the Control Engineer should be encouraged to take advantage of the unique capabilities of man, to enrich man's role in the system. An important objective of the system should be greater humanization and opportunity for human self-actualization and growth.

5. The Control Engineer should give serious consideration to re-oreinting or reshaping his profession and its educational base to include exposure to economic, social and psychological factors; failure to incorporate such aspects in his thinking and activity will severely limit the effectiveness of his designs.

At a subsequent committee it was agreed to follow up the Workshop by in-depth investigations by the Committee at a number of factory locations. It was also agreed to propagate the work of the Committee by means of a Newsletter. It was further decided to make preparations for the August 1975 IFAC Congress at Boston, U.S.A. and it was agreed to arrange another Workshop. At this time the Chairmanship of our Committee passed from Phil Sprague to Fred Margulies.

NEWSLETTERS

As this report is being written, the Committee's third newsletter is due and plans exist for editing the fourth and fifth newsletters.

The procedure is that responsibility for each newsletter rests with a named individual, as follows:

No. 1 – H. Rosenbrock
No. 2 – J. Dockstader
No. 3 – A. B. Aune
No. 4 – P. Schuh
No. 5 – S. Aida

The master copies are sent to a number of committee members for duplication and distribution on an area basis. Distribution is undertaken by the following:

N. & S. America – J. Dockstader
U.K. & TC 9 – H. Rosenbrock
West Germany – P. Schuh
France & Switzerland – F. Muller
Austria – F. Marquilies
Scandinavia – A. B. Aune
Socialist Countries – N. S. Rajbman

The committee's Newsletters aim to provide:

– Preview of forthcoming events
– Reports on conferences, workshops, meetings etc.
– Reviews of books and papers
– outstanding papers in full
– reports on research
– suggestions for committee work

The first Newsletter was issued in March 1976 and the second in November 1976. The third Newsletter is due in April 1977. It is primarily aimed at meeting the needs of committee members and their friends and an initial circulation of around 100 copies was envisaged. In practice many more copies than this have been circulated.

Reciprocal arrangements regarding the circulation of New Newsletters have been made with Council on Quality of Working Life through the Chairman, Professor A. T. M. Wilson.

Similar arrangements for exchange of Newsletters have been made with the IFIP committee TC 9 devoted to the study of relationships between computer technology and society.

FACTORY VISITS

At the Bad Boll Workshop, proceedings fell under three main headings:

– process industries
– manufacturing industries
– office work

The initial factory visit was particularly relevant to the first of these. The 'Hoogovens' plant at Ijmuiden in the Netherlands is one of the most up-to-date steel works in Europe with a high level of automation throughout. It has two hot strip mills, one of which is equipped with a comprehensive computer control system. At the time of the Committee's visit (February 1976) the system had been in operation for some years and there had just been concluded an eighteen month study into Human Factors by a co-operative team from Hoogovens staff, the University of Delft, and the British Steel Corporation. The study had been financed with assistance by the European Coal and Steel Community. The meeting was attended by fourteen committee members who spent two days visiting and discussing the plant together with senior staff from Hoogovens and the co-operative study team.

Participants expressed their appreciation of the character of the meeting and it was agreed to repeat the structure in future.

Full reports of the meeting appeared in Newsletter No. 1. A main impression for the meeting was the considerable attempt being made to bring together working experiences of operators, behavioural scientists, engineers and managers in an attempt to make improvements for the future. Another was the organisation which exhibits a very high standard of plant technology and which attempts to involve its work force in an open exchange of information. Thirdly, it was interesting to see the attempt of maintaining good visual and physical proximity to the process, with the control room used as a meeting place and management centre.

It had been hoped to compare conditions in the highly automated hot strip mill No. 2 with those in the less advanced mill No. 1. This proved impracticable for a number of reasons. Whilst there was a cost benefit to Hoogoven in terms of improved heat economy and production quality there were no labour savings: in fact twelve relatively unskilled people were replaced by a need for an equivalent number given with higher levels of skill and operator training in a lengthy process.

Our second factory visit was the Olivetti's main works at Ivrea in Northern Italy. Olivetti face particularly challenging problems because of the change in their technology base from electro-mechanical 'fein werk technik' to electronics.

At the time of this visit the committee comprised forty-two members from seventeen countries and thirteen members were in attendance from ten countries. Two days were spent on the visit to Olivetti and two further half-days were spent discussing committee business.

Olivetti have made significant progress in two directions in response to the difficulties human beings face when coupled to mechanical flow lines and asked to work repetitively on short cycle times. Firstly they have been introducing production systems based on group technology since 1963. In 1976, 2,000 people were so organised at Ivrea. Secondly, since 1972 they have been introducing robots; in 1976 they had twenty at work, equivalent in speed to some sixty men, and able to work more or less continuously.

There were intensive discussions during our visit both about Group Technology and about Robots. There is a full report on our visit in Newsletter No. 3. Among many impressions of our visit was the big investments being made by Olivetti both in its people by means of training schemes and in its technology. The effects on the structure of work and upon the management structure were considerable with much evidence of matrix and project management systems.

Visit to Daimler-Benz, Sindelfingen (FRG)

In amy 1977, a visit was paid to the Daimler-Benz plant at Sindelfingen, where Mercedes cars are being assembled. About 10 committee members participated.

The daily production is about 1500 cars, of which are sold to employees for a reduced price. The total working force is 34000, including 10000 guest-labourers.

Job enlargement has been introduced in order to improve the cycle time in some monotonous jobs; for instance a worker will take every fourth car and spent 4 x 2 minutes on it. Still there are several people strongly tied to the pace of the machine, amongst others in the pressing of body parts and in the painting shops. The creation of more favourable working conditions tends to enlarge intermediate buffers and to decrease productivity.

Many robots are being used, mainly for spot welding and painting. However, locations which are difficult to reach remain to be done by human beings.

Visit to Hewlett-Packard, Boblingen (FRG)

Also in May 1977, a visit was paid to the Hewlett-Packard plant at Boblingen. Of the total working force of 1100, 380 are employed in production. Small series of a great variety of electronic instruments are being manufactured.

The interior of the buildings has been arranged in open style, where even the department head's desk can be seen by everybody.

The organisation is based on the principle of 'management by objectives". Every year objectives are formulated by the Supervisor under negotiation with the person concerned. The latter works independently, with a semi-annual inspection by the supervisor. On the lowest level however, there are difficulties in defining feasible objectives. Special attention should be paid to the training of foremen in applying the principle on their level.

Job enlargement has been realised in the assembly of printed circuit boards. The testing of these boards will be done by computer, which enables job upgrading for quality inspectors to programming, or to customer servicing.

General Impressions from the Factory Visits

There is a variety of reasons for starting projects on humanisation of work, such as: High absenteeism and turn turn-over, higher levels of education of young employees, changes in technology accompanied by changes in jobs, pressures from trade unions, new legislation, and/or initiatives by management. In any case, such projects only have a chance of success if they are supported by management.

An important condition imposed upon management, particularly under the present economic conditions, is to keep production efficiency at least on the same level. Additional costs of humanisation of work must therefore be compensated by some form of productivity improvement. Giving more individual responsibility for product quality, or better information to the operators about process performance can help secure this end.

Great attention should be paid to training, not only of the workers concerned, but also of their supervisors. In advanced automation projects, the operators should understand the decisions taken by the computer.

Modern computer and communication technology can enable the worker to take actions which traditionally are the realm of the supervisory level. This, inevitably, must lead to changes in the organisation structure.

The present generation of industrial robots have several limitations, hence a certain number of monotonous tasks still have to be allocated to human beings.

IFAC Congresses

The International Programme Committee of the IFAC Congress 1975, which was held at Boston (U.S.A.), agreed to include a specially commissioned paper among its list of plenary sessions. In fact a poll put the Social Effects of Automation fourth on a list of sixteen titles for choice.

This paper was a joint effort by K. Bibby, F. Margulies, J. Rijnsdorp and J. Withers, and presented by J. Rijnsdorp. In addition, two special interest sessions were organised comprising twelve papers which stimulated great interest and lively discussion. A well-attended round table meeting under the chairmanship of Dr. Ellis-Scott provided further suggestions concerning the future activities of the committee.

For the coming IFAC Congress, which will be held in Helsinki, June 1978, the following survey papers have been proposed:

- Technological change, productivity and employment, by Mr. Cooley and J. Withers;
- Co-operation between control engineers, Social scientists and users in automation system design, by Aune, M. Cooley and J. Rijnsdorp.

Further, round table discussions will be held about these two subject, and a third one will be devoted to Man-Machine Interfaces for Process Control in cooperation with Technical Committee 6 of "Purdue Europe"

Finally, two technical sessions are envisaged about Social Effects of Automation in the Printing Industry, to be organised by Kavonius and Withers; and about Automation in Offices, to be organised by Bjorn-Andersen.

WORKSHOP AT ENSCHEDE (Netherlands)

This paper is intended for presentation and discussion at the Workshop about Case Studies in Automation, related to Humanization of work, to be held at Enschede (Netherlands) from 31 Oct. to 4 Nov. 1977.

A provisional programme was agreed to give three working sessions or case studies, with two sessions devoted to a summary and recommendations, and hopefully, a technical visit to Philips.

The workshop is being organised by a local committee led by Professor J. Rijnsdorp.

FUTURE WORK

The committee meets very infrequently and its membership is spread across many countries and is composed of individuals with heavy committments.

It has therefore been necessary to concentrate on a relatively narrow front but with a strong theme in order to achieve success.

From the foregoing it will be appreciated that the Committee's past activities have focussed particularly on the problems of humanising work in manufacturing situations with special reference to automation. It is felt that this concentration has yielded good results.

Our factory visits, newsletters, workshop programme, and congress contributions have all yielded favourable comment and enabled many to benefit for the sharing of working experience.

Whilst however the committee feels that the selection of the manufacturing sector is justified by the wealth it so necessarily creates for society to spend, it has to acknowledged that there is an increasing employment opportunity in the service sector and that other aspects and activities such as medical aid, transport and communications all deserve study. The possible scope of the Committee's activities is therefore very wide and it is planned to hold discussions at Twente to see what more might be fruitfully explored in the future.

2. SUPERVISION OF PROCESSES AND SYSTEMS

HUMAN CONTROL TASKS: A COMPARATIVE STUDY IN DIFFERENT MAN—MACHINE SYSTEMS

C.L. Ekkers, C.K. Pasmooij, A.A.F. Brouwers and A.J. Janusch

Netherlands Institute for Preventive Medicine, Postbox 124, Leiden, The Netherlands

1. *Introduction*

Technological and economical developments in recent decades have affected in a number of ways the characteristics of industrial production systems, transport and shipment systems, and information processing systems. Product diversification and the necessity of a keen control of product costs made advance and have led to production on enlarged scale. At the same time the systems have grown in complexity which had to do with the development of improved control devices and automated information processing equipment.

Extra dimensions have been added to these developments by the introduction of the computer: fast processing of huge amounts of data and automatic control of complex systems.

Examples of the above indicated technological process of growth, generally referred to as "automation", can be found in the chemical industry, electricity-works, traffic control systems, and in computerized administrative systems.

As the examples indicate "automation" is being applied in a variety of settings. Nevertheless, due to the general characteristics of automation, these systems do have a number of common features:
. system and/or subsystem operations are to a lesser or greater extent automatically executed,
. information presentation with regard to the state and time-history of the process and system-components is centralized and displayed by means of an interface,
. this interface in most cases also contains control devices,
. the human operator can either adjust the controlled variables directly, or has the possibility to change setpoints of the automatically controlling system.

In spite of the fact that many of these systems are highly automated, the human operator still has an important function in the man-machine system. Especially in those situations when the automatic controller fails, the contribution of the human operator is of crucial importance.

The tasks the operators in these systems have to carry out are alike in a number of respects as a consequence of the fact that these systems share a number of common features.

In most automated man-machine systems it can be seen that a considerable amount of time is spent by the operator on monitoring the system. We call this the monitoring aspect of the task.

Generally less time (depending on the degree of automation) is spent on control activities. These control activities may be related to adjustments of the production process, as well as to reactions with regard to disturbances in the system. We call these activities the control aspects of the task.

Often a large amount of time is spent on communication activities. We can make a distinction here between consultation with colleague operators or the shift leader on the one hand, and indirect control of the system by means of orders to the field operators on the other hand. We call this the communication aspect of the task.

At last there are a number of activities which are aimed at the future effectiveness of the system such as administration and operations concerned with the maintenance of the system. We call these the future effectiveness aspects of the task.

What kind of factors do determine the more specific content of the task and the distribution over time of the different activities of the human operator? It can be stated that a man-machine system as a whole is the final result of a design process in which different functions are allocated to either the machine or man. However generally spoken this allocation process is not so much based on criteria of human or social origin, but more on technological and economical criteria.

It follows that between technical equipment and the requested human contribution to system performance a number of potential discrepancies can arise. For instance with regard to monitoring, problems of maintaining vigilance can exist. The consequences of misjudgments of an evaluative or predictive nature can be considerable. The maintaining of a high level of control skill in situations that require few interventions is to some extent irreconcilable.

Rigidity of control-procedures, framed in instructions, can be inappropriate in critical and emergency situations. In these situations overload or high levels of stress may occur for the operator.

Superimposed on aspects of optimizing the fit between man and machine, are a number of possible discrepancies of an organizational nature. For reasons of productivity automation was mostly accompanied by the introduction of shiftwork, and sometimes resulted in a reduction of the number of operators and thus of social contact.

When such discrepancies exist, these may lead to a decrease in task performance and possibly to injuries of the system. On the other hand they may lead to negative effects for the operator himself such as boredom, or sometimes overload of the operator, and subsequently to lack of motivation, absenteeism and negative effects on health.

In short job satisfaction, mental and physical health of the operator can be affected by a number of factors, which can be embedded in the technical system, task structure, task-fulfillment, and the organization. To clear these relationships is the aim of the present study.

2. *Some considerations with respect to possible research*

It has been stated that potential man-machine problems basically are born in the design phase of the construction process. It should be emphasized that research must aim at the development of design criteria which can be applied to this construction process. As mentioned in the introduction, in most man-machine systems one can make a distinction between the technical system the human operator has to work with, the task he has to perform, and the organization in which he is functioning. Effects on job satisfaction and health of the human operator are assumed to originate from discrepancies between demands from the technical system, the task, and organization on the one hand, and the capacities, skills, and needs of the human operator on the other hand.

What we need are design criteria based on knowledge about these discrepancies, which can be related to objective properties of the technical system, task and organization. What kind of research is needed to deliver these design criteria? The method for which we have chosen is to compare a number of different task situations with regard to aspects of the technical system, the task, and the organization on the one hand and job satisfaction and health on the other hand. Rather than a thorough study in one particular task situation, which is an approach more frequently chosen, this type of research may lead to results which can be applied also in other situations than the ones investigated.

Research in the field of man-machine systems has started in earlier days with a monodisciplinary viewpoint, originating in the medical, social or technical sciences. Consequently these studies have been focussed on one particular part of the problem field (e.g. Evrard 1973, Bainbridge et al. 1968, Kuprijanow 1969). More recent studies show the development of multidisciplinary teams which cover more aspects of the man-machine area (EGKS study 1976, Vogt et al. 1977). Both approaches have been followed in the laboratory environment, as well as in the field situation. In the laboratory situation more fundamental problems can be answered, whereas field study involves the typical aspects of the real life situation. However, research of the latter type in general has been limited to an investigation of a small number of task situations simultaneously. The gain in depth in most cases means a loss of relevance for general application, because a small number of task situations can lead to results which are biased by typical characteristics of the situation where the investigation takes place.

In literature no extensive multidisciplinary study can be found, in which comparisons are made between a great number of completely different man-machine systems. The present research will follow such an approach and in order to get a reasonable variance on the different factors in the technical system, task and organization a total number of 25 task situations are selected. This implies that the unit of the comparative analysis will be the man-machine system (with the data of individual operators in that task situation aggregated for instance by means of the arithmetic mean of the data) and not, such as is the case in many field studies on one or a few systems, the individual operator within the system. In our view data of the latter kind can never lead to more general applicable research results which provide design criteria because they only reflect differences in individual reactions on one task situation, while the objective characteristics of the situation are not varied. Because design criteria must refer to these objective characteristics variation in these characteristics is necessary to determine their effects on the human operator. In the laboratory this is being done by systematically varying a small number of factors in the system. In the field study we make use of the "naturally" occuring differences between a number of man-machine systems, which makes possible an evaluation of the effect of more factors than in a laboratory situation, however in a less exact manner. Eventually results of both approaches must be combined to provide optimal applicable design recommendations.

An important assumption underlying this approach is, that groups of operators in the different man-machine systems one wants to compare do not show great a priori differences on a number of characteristics, so that it is not possible to attribute differences found between different groups of operators either to characteristics of the different man-machine system or to these a priori differences. This has been tested in a pilot study in four task situations, which preceded the present research and there was evidence that most differences between groups of operators could be attributed to system characteristics and not to a priori differences (Ekkers et al. 1975). As a result it can be stated that in investigating a great number of different task situations subjective data (e.g. interviews) concerning groups of operators can be used to make statements about effects of objective system characteristics.

As stated in the introduction, automation has made advance in many different fields of industry. While these man-machine systems can be completely different in terms of the technical system (production process, capital invested, dimensional scale, process dynamics) and organization (shift system, number of operators per shift, type of leadership etc.) they still provide tasks for the human operator which have a number of elements in common. This means that in selecting the tasks, we can choose from a great number of completely divergent man-machine systems.

In order to avoid however that conclusions and recommendations which will result from this study should become of too general a character, it is necessary that the tasks under investigation do not diverge too much. Therefore two criteria have been formulated which these tasks have to meet:
. the human operator has to monitor and/or control a more or less automated technical system (panel operator),
. all information input and control output of the panel operator has to be concentrated in one central control room.

It has been possible to select 25 task situations which meet these criteria. These task situations can be classified with respect to the type of production process in the following groups: chemical industry, power generation plants, traffic control systems, computerized administrative systems, and various other systems.

3. *Description of the comparative study*

The data that are gathered for this investigation can be roughly divided with respect to the method of data collection into three groups. First the data that are obtained by means of structured interviews with the operators. These data are concerned with the perception and appraisal by the operators of aspects with regard to the technical system and the task of the operator. Furthermore the operators have to fill out a questionnaire concerning organizational aspects, job satisfaction and subjective health. Sets of questions from these interviews and questionnaires are used for the construction of dimensions which characterize each task situation.

A second group is formed by data which are gathered at the management level. A structured interview with a design engineer provides information about the technical system. The same procedure holds for task and organization, though this information is provided by the personnel department. At the same time the personnel department supplies figures with respect to sickness absenteeism of the operators under investigation.

Finally the last group of data exists of audio-visual recordings with a duration of 30 minutes of task execution per period. Together with these recordings electrocardiographical data of the operators are registrated. The number of registrations per task situation comes up to approximately 25, proportionally distributed over the different shifts. This group of data will be used in determining the activity pattern of the operators, and in estimating their workload.

At this very moment the phase of gathering the data has been terminated. The analysis of the data is in full progress. The publications can be expected at the end of this year. In this paper we will present some preliminary results obtained in 8 different task situations, which will be described here.

1. Dredge system. Hopper dredge ships have to ensure sufficient depth in some particular fairways. For this purpose these ships are equiped with two suction pipes, each controlled by an operator. These operators have to take care of the pump systems on board as well.

2. Ship traffic control system. This system safeguards a secure and efficient ship traffic in a fairway. The movements of the vessels are observed by means of radar equipment by the operators, who provide these vessels with information by radio.

3. Experimental electricity works. This system is set up with the aim of testing sodium components under fully operational conditions. The operator has to control the system during the experiments.

4. Computerized administrative system. The job stream which the computer has to execute is under supervision of an operator. Within the framework of an optimal use of the computing facilities, he can change the priority codes of the system.

5. Underground traffic system. Maximal safety is guaranteed by an automatic anti-collission system. The operator has to supervise this system, and if necessary to set traffic lanes manually.

6. Conventional electricity works: steam supply. This part of the system provides the turbines with steam of a demanded quantity and quality. The operator has to control the steam generator for this purpose.

7. Conventional electricity works: power generation and distribution. The turbines are centrally switched on or off, according to the demand of the local area, or the national grid system. Control is executed by the computer and is supervised by the operator.

8. Blast furnace. The furnace process, as well as the auxiliary systems are centrally controlled. The operator has to supervise the operations of the system.

For these 8 task situations the above described data have been obtained. It will be described more specifically in the following which methods have been used in order to make these data applicable for the analysis at the level of the task situation. This will be done respectively for the technical system, the task, the organization and the factors concerned with job satisfaction and health.

3.1. *The technical system*

With the help of the data obtained in the interview with a member of the technical management, a number of characteristics of the plant hardware, the control system and the man-machine interface is formulated. These characteristics are focussed on aspects as the degree of automation, process dynamics, batch or continuous production, capital invested, system ergonomics, etc.

The interviews with each of the operators participating in the investigation deliver subjective data with respect to the three sub-systems mentioned. Sets of questions are determined, and the arithmetic mean of the scores on these questions over all operators per task situation is computed. These sets of questions each represent a specific dimension of the technical system as it is perceived and valuated by the operators. These dimensions of the technical system are the following:

. Complexity (COMP). This dimension incorporates the number of controlled process variables and the degree of mutual coherence, the amount of information presented to the operator, etc. A high value signifies a high degree of complexity of the system.

. Degree of automation (AUT). This dimension represents the operator's opinion with respect to the degree of automation of the system during starting-up, stationary and shutting-down conditions. Data with respect to the allocation of control actions between man and automatic controller are involved as well. A high score points to a high degree of automation.

. Process dynamics (DYN). Dynamical properties of the process under control and of system responses to control actions of the operator, as well as the information presentation rate, are represented in this dimension. A high score points to relatively fast operations.

. Sensitivity of the system to disturbances (DIST). This dimension is compiled of data with regard to the occurrence of disturbances in the plant hardware, the control system and the man-machine interface. More frequent and extended disturbances will lead to a high score.

. Controllability of the process (CONT). The adequacy of the control system, as well as the subjective feeling of the operator of being able to control the system manually, contributes to this dimension. A high score points to an adequate controllability of the system.

. System ergonomics (ERG). Represented in this dimension are ergonomic aspects related to information presentation and control equipment, as well as the overall lay-out of the man-machine interface. A high score means an ergonomic sound interface hardware.

. Task environment (TEN). A high score means a positive perception of aspects of the task environment, e.g. noise, illumination, temperature, etc.

. Indirect information intake and system control (IND). This dimension quantifies the information intake and control actions by the operator which do not take place by means of the interface. A high score stands for frequent actions of the operator without the use of the interface e.g. giving orders to field operators.

3.2. *The task*

A description of the task of the operator is set up with the aid of a job description obtained from the personnel department, and interviews with operators as well as observations in the control room. The description will be focussed on the functional aspects of the activities of the operator.

As mentioned additional audio-visual recordings are made of the task fulfillment of the operators. These systematic observations will be used to determine the amount of time the operators spend in each task situation on the four already mentioned categories of the task (see introduction): monitoring, control, communication and future effectiveness. Together with the data computed from the electrocardiographic recordings, these data will serve in the estimation of the level of workload. This will be subject of forthcoming publications. Subjective data with respect to various aspects of task assignment and task execution are collected by means of the structured interviews with the operators. The same procedure as explained under 3.1. has been applied in order to construct the following dimensions with respects to the task:

. Detailedness of instructions for control of the system (DETSYS). A high score means that these instructions are accurately described for different situations that can occur.

. Detailedness of instructions with regard to other activities (DETOTH). A high score means that the way in which these activities should be executed by the operator, are accurately described.

. Formal or informal role of the instructions (FOR). A high score means that the instructions are formally treated in the work situation.

. Acceptation of instructions for control activities (ACSYS). Positive valuation of the presence and content of instructions with respect to the control activities to be performed by the operators, is represented in a high value of this dimension.

. Acceptation of instructions for other activities (ACOTH). The same holds with respect to other activities of the operators.

. Activities (ACT). A valuation of the time which the operator can spend to various activities as well as an appraisal of the activity pattern is represented in this dimension. A positive valuation leads to a high score.

. Information complexity (INF). In this dimension the extent in which the operator has to take into consideration a variety of information sources during disturbances, is represented. A high score points to a strong coherence between the various sources of information.

. Work procedures (PROC). This dimension represents in what extent the actual task execution is prescribed by procedures. Fixed procedures will lead to less degrees of freedom for the operator, and to a high value of this score.

. Control uncertainty (UNCER). Possible problems which the operator can have in selecting the appropriate control actions is ment here. Many difficulties lead to a high score.

. Load due to disturbances (LODIS). A high value of this dimension represents a high perceived level of load, in the case of the occurrence of disturbances.

3.3. *Organizational aspects*

Some formal characteristics of the organizational system are obtained by means of the already mentioned interview with a member of the management. Additional information is obtained by observation on the spot, and this is also incorporated in the description. The valuation of the operators with regard to these aspects are obtained by means of the questionnaire.

In the scope of this paper a short enumeration of the factors which can be distinguished with respect to the organizational system will be sufficient:
. shift system
. safety aspects
. economical aspects
. responsibilities
. division of tasks, and task roulation within the group
. relations with colleagues
. style of leadership.

3.4. *Job satisfaction and subjective health*

By means of a questionnaire filled out by the operators data were obtained with regard to different aspects of the task which are related to job satisfaction. The aspects which will be treated in this paper are: experienced stress in the work, experienced work load, amount of variation the operator experiences in his task, and feelings of achievement as a result of his work. Stress refers to experienced tension in the work, work load to the amount of work. On a number of questions the operator furthermore indicated how he felt about his health. Three subscales are to be distinguished. The first contained positive statements about his health, with which he could agree or disagree (e.g. "I am in a good physical condition"). We call this scale positive subjective health. The second contained negative statements (e.g. "I don't think I am in as good a condition as other people of my age"). We call this scale negative subjective health. We expect a correlation between these subscales, but not very high, because the absence of complaints does not necessarily mean that one experiences a good state of health. The third subscale is developed by Dirken (1967) and gives an indication of the number of psychosomatic complaints the operator has. This scale is expected to be positively correlated with negative subjective health.

4. *Results*

As an illustration of the approach we have chosen for research in the man-machine systems area we will present now some results of the analysis of the data from the first 8 task situations. This analysis will be restricted to the questionnaire data from the operator with regard to the technical system, the task, job satisfaction and subjective health. We will be dealing with the data of 81 operators distributed over the 8 task situations mentioned. As we have explained already, the analysis will not be on the level of the individual operator, but on the level of the task situation (N=8). Eventually this kind of analysis will be possible over 25 task situations, so the results that are reported here have a preliminary character.

For this analysis we use for each task situation the arithmetic means of the scores of the individual operators in that task situation. These scores are computed for each of the factors related to the technical system, the task, job satisfaction and subjective health that are mentioned in the previous paragraph. The relationships between these factors will be checked by means of Pearson's correlation coefficients. As a consequence of the limited number of tasks, the level of significance (5%) will only be reached for coefficients $\geq |.62|$. Only these correlation coefficients will be taken into consideration, although sometimes somewhat smaller coefficients will also be mentioned.

As a simple model of analysis we treat the factors related to the technical system and the task as independent variables and job satisfaction and subjective health as dependent variables with sometimes job satisfaction as an intervening variable between task or technical system on the one hand and subjective health on the other hand. In later analyses factors with regard to task fulfillment and work load as obtained from observational and physiological data will also be taken into consideration. It must be noted that, strictly spoken, in a correlational analysis, it is not possible to conclude for causal relationships.

Starting with the analysis within the dependent variables we see a negative correlation between positive and negative subjective health (r= -.60) and a positive correlation between negative subjective health and number of psychosomatic complaints (r= .61). The absence of a correlation between psychosomatic complaints and positive subjective health is in accordance with the assumption mentioned before, that the absence of complaints is not the same as a state of positive health. Therefore we will treat relationships with these factors separately.

For the four job satisfaction variables we find a significant positive correlation between experienced stress and work load (r= .64) and one between feelings of achievement and experienced variation in the work (r= .68). Work load shows also rather high, but non-significant correlations with achievement (r = .57) and variation (r= .51). As we would expect feelings of achievement are negatively correlated with negative subjective health (r = -.83) and psychosomatic complaints (r = -.65). Also the amount of variation in the work is negatively related with negative subjective health (r= -.65). However also the subjective work load is negatively correlated with negative subjective health (r= -.87) and the same tendency is found for experienced stress in the work (r= .59 with positive subjective health and r = -.48 with negative subjective health). From this result it may be concluded, that in the population investigated experienced work load and stress have a positive effect on the subjective health of the operators. This would be in accordance with the more often postulated hypothesis, that in automated man-machine systems underload and too low levels of stress are a more serious problem than overload and too much stress (Bibby et al. 1975).

A next step in the analysis is to find out which dimensions in the technical system and the task have a relationship with the job satisfaction variables and subjective health. We start with a description of the dimensions that affect subjective work load and stress together with subjective health. The relationships between these variables are schematically represented in figure 1.

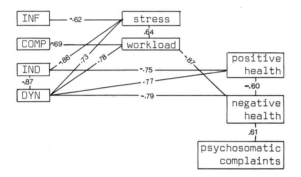

Figure 1: schematic representation of the relationships between dimensions of the technical system and the task, and job satisfaction and subjective health.

It appears, that the factors indirect control (IND) and information complexity (INF) have negative relationships with stress (r= -.86 and r= -.62 respectively), whereas the factor process dynamics (DYN) has positive effects on stress (r= .73) and load (r= .78). The factor process dynamics has also a relationship with positive (r= .77) and negative (r= -.79) subjective health. This applies also for the factor indirect control as far as the relationship with positive subjective health is concerned (r= -.75). Also it should be mentioned that the factor indirect control is negatively correlated with the factor process dynamics (r= -.87). The factor work load is also negatively influenced by the perceived complexity (COMP) of the system (r= -.69).

One might conclude from these relationships that in systems with a high degree of complexity (of the technical system and/or the information presented to the operator) and a relatively slow process combined with a large amount of indirect control, low values are found for experienced stress and work load which are also related with feelings of a less good health.

Other interesting relationships are found between the technical system and task dimensions and the amount of achievement the operators experience from their work.

The factor achievement in the job satisfaction questionnaire is positively related with the dimension activities (ACT), controllability of the process (CONT) and system ergonomics (ERG) (r= .76, r= .79 and r= .62 respectively). The dimension CONT has also a relation with positive (r= .64) and negative subjective health (r= -.64). It thus seems that good ergonomic conditions (related to the system), a system which is easy to control and the availability of a meaningful pattern of activities are good preconditions for feelings of achievement and consequently of health of the operator.

At last we will mention some relationships between technical system and task dimensions and subjective health without accompanying correlations with job satisfaction.
It appears that in systems where the sensitivity for disturbances (DIST) is high, positive subjective health is low (r= -.62). Although a high score on this factor implies also a somewhat higher variation in the work (r= .56), an increase in disturbances does not seem a desirable solution.
Furthermore we find a relationship between good working conditions (TEN) and the number of psychosomatic complaints (r= -.72). Although it is often claimed, that physical working conditions (temperature, noise, dust etc.) are not so much a problem in automated man-machine systems, this result indicates that improving these conditions may have prosperous effects on the operators.

The importance of the detailedness of the system prescriptions (DETSYS) and the acceptation of these prescriptions (ACSYS) by the operators is indicated by the positive correlations of these factors with subjective positive health (ACSYS: r= .76), subjective negative health (ACSYS: r=-.72) and number of psychosomatic complaints (DETSYS: r= -.88). A better insight in the nature of the relationships would be obtained, if one would use multivariate methods of analysis (partial and multiple correlation coefficients). However the limited number of cases on which such an analysis would be based would leave so small a number of degrees of freedom, that only very high coefficients would be significant. Such analyses will be made in the total analysis over 25 cases.

5. *Conclusion*

Although the present analysis is restricted to 8 task situations, it is possible to draw some tentative conclusions. It appears that the subjective work load of the operator is an important variable, that is related with technical system and task dimensions as well as with subjective health. A low subjective work load is found in large complex systems with a lot of indirect control (contact with field operators). This low level of subjective work load of the panel operator is accompanied with a low level of subjective health. It is obvious that a remedy can be found in making available for the operator more meaningful activities, which has - as appears in this study - a positive relation with feelings of achievement in the work. Also it is clear however, that supplying these activities must not lead to a diminution of the controllability of the process or to a greater sensitivity for disturbances of the process. An important limitation of this study is the fact, that all variables used in the analysis are based on subjective experiences of the operators. With regard to health and job satisfaction this is not so much a problem, because these variables are not measurable in a strictly objective way. As for system and task characteristics however one would like to have also more objective data to make design recommendations easier to deal with. Therefore in later analysis also objective data with regard to the technical system, the task and the organisation of the work will be incorporated in the analysis as well as data based on behavioural observations and physiological measurements of the operator during task execution.

The present analysis may lead to the conclusion that a comparative study of quite different man-machine systems, which implies an analysis on the level of the system and not on that of the individual operator, can provide meaningful results with regard to the human aspects of man-machine systems.

6. *References*

L. Bainbridge, J. Beishon, J.H. Hemming and M. Splaine, A study of real-time human decisionmaking using a plant simulator, Oper. Res. Quart. 19, 91 (1968).

K.S. Bibby, F. Margulies, J.E. Rijnsdorp, R.M.J. Withers and I.M. Makarow, Man's role in control systems, IFAC Congress (1975).

J.M. Dirken (1967) Arbeid en stress; het vaststellen van aanpassingsproblemen in werksituaties, Wolters - Noordhoff, Groningen.

C.L. Ekkers, W.T.M. Ooijendijk en J.J. Schwarz, Menselijke stuur- en regeltaken; verslag van een onderzoek bij vier verschillende stuur- en regeltaken, Leiden NIPG/TNO, (1975).

Ergonomic evaluation hot stripp mill 2 Hoogovens, EGKS study No. 6245/12/804-604, (1976).

E. Evrard, Problèmes médicaux relatifs au personnel chargé du contrôle de la circulation aérienne, XXIéme congrès intern. de médicine aéron. et spatiale, (1973).

A. Kuprijanow, The effect of airspace loading by controlled air traffic in the air traffic control system, Proc. of the IEEE 58, 381 (1969).

J.J. Vogt, R. Foehr, E. Kuntzinger, L. Seywert, J.P. Litert, V. Candas and Th. Van Peteghem, Improvement of the working conditions at blast-furnaces, Ergonomics 20, 167 (1977).

BASIC TRANSFORMATIONS: THE KEY POINTS IN THE PRODUCTION PROCESS

Christian Schumacher

British Steel Corporation, 33 Grosvenor Place, London SW1X 7JG, England

At its most fundamental, work is concerned with the transformation of nature by man, with changing raw materials into finished products; and industrial work is the rationalisation of this process by applying scientific laws and techniques to it which speed it up and make it more efficient. A modern industrial production system takes in raw or semi-finished materials and seeks speedily, accurately and cheaply to transform them into a finished product. Each activity in a production process is a step on the route from raw nature to a useable item or commodity. Its primary function is change.

The idea of a production system as a change process is a broad one; it embraces extraction and agriculture - mining, harvesting, the business of wrenching raw materials into the area of man's attention. It embraces manufacturing - the business of transforming the materials of raw nature into useful goods for man. And it embraces services - from dry cleaning to machine maintenance to computer services - that manipulate natural laws and materials to man's needs for order and satisfaction.

The place where we can most readily observe 'change' in a production system is within the individual machine. This is where the process of change is centred, where the natural laws of physics and chemistry that man has so far discovered have largely to be applied. Every material has certain physical, chemical, metallurgical or biological properties which determine the conditions under which it can be changed from one state into another. In order for these changes to take place, certain pressures, chemical or electrical conditions have to exist and certain agents or ingredients have to be present. The demands are basic and apply whether the product is aluminium foil or petrol for the family car.

It is up to the scientists to discover the natural laws which govern the change process and it is up to the engineers to design them into the pieces of machinery upon which production systems are based. However, whatever the natural law and whatever the make of machinery employed the notion of 'change' is common to all, and underlies the activities of all industries and of whatever kind.

When one perceives work as a change process it is possible to distinguish different degrees of change in the different activities which make up a production system. By following each stage in the production process, one can categorise how much change takes place to the product at each point.

At the lowest level of change are storage activities. Here nothing changes, except time. The product just lies there, as raw materials on a stockpile, coal in the seam, a component on the store shelf, wine in a bottle, papers in the filing cabinet. All extraction, manufacturing and service industries have their share of storage activities in almost every department; they are called stores, stocks, reserves, or buffer commodities.

At the next lowest level of change are <u>transportation</u> activities, where the product is moved from one place to another. Again, nothing happens to the internal constitution of the product, but in addition to changes in time there is also a change in place. Transportation activities, too, occur in every production system: whether the product is conveyed by lorry, railway wagon, forklift truck, conveyor belt, hoist, skip, pipe, crane or even by human beings moving it manually from one workbench to another, as it travels on its journey towards completion and sale.

Then come the types of change in which the product itself is altered. These are clearly more fundamental to production than those changes in which the product is simply moved or stored - but itself is left intact. Three main levels of 'product change' can be distinguished, in ascending order of significance. First, there are those changes which only affect the external appearance of the product, ie peripheral or cosmetic changes. Such things as wrapping, packing, polishing, trimming, cleaning, sandblasting, grinding, chipping, sorting, counting, filtering, folding, fall into this category of change. Here, nothing happens to the substantive core of the product - that remains as it was - but the external appearance is altered in some way. Usually, these minor, or <u>ancillary operations</u> take place both at the preparatory and at the finishing stages of production, that is at the input and output stages.

Secondly, there are those changes which significantly affect the internal constitution of the product itself, but which do not themselves actually constitute the major conversion activity as such. They are usually associated with it, however, are adjacent to it and supportive of it in the sense that they involve an essential mixing of ingredients prior to or following the major change activity itself. Splitting, kneading, cutting, gluing, aggregating, separating, welding, binding, nailing, coating, are examples. These supporting activities may be called <u>supplementary transformations</u> since they are necessary prerequisites for the major change activities. They exist for these major change points rather than the other way around.

Finally, we come to the major conversion activities themselves. These are where the major physical, chemical, electrical, biological or functional changes take place in the product itself; where, so to speak, the 'whole' becomes greater than the parts. These are the points where the iron is turned into steel, the clay becomes a pot, the components become an engine, the statistics become a mathematical solution, or the product becomes useable by the consumer at the point of sale. They are the central or 'key' points in the production process, the points where the actual transformation of nature takes place in its most complete form. We have called them <u>basic transformation</u> activities and one can find them in every conceivable kind of work, under every kind of guise: baking, melting, refining, pulping, mixing, firing, igniting, cooking, fusing, stamping, assembling, solving, selling, transmuting, magnetising, extracting, inducing, reducing, rolling, synthesising - all these and many others.

In a 'hierarchy of change' basic transformations are the 'peaks' and storage activities are the 'troughs'. All processes can be described in this way - an orchestrated harmony of high notes and low keys around a theme of change.

Here are some typical examples of productive processes displayed as change hierarchies taken from industry.

Degrees of Change in Production
Different Technologies

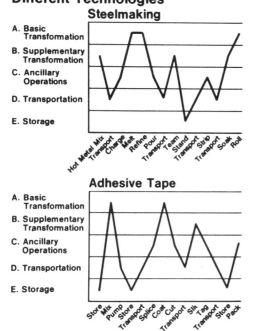

As can be seen, the change profile for each different process is unique to that process (for example Steelmaking & Rolling has only two basic transformations, while Porcelain making has four), but all exhibit the same change tendencies, and so can be compared with one another.

This way of describing a production process allows us to make some interesting observations about the individual activities which comprise it. Significant differences may be discovered between the characteristics of activities lower in the hierarchy and those higher in the hierarchy.

In the first place in a basic transformation there is generally a greater complexity of factors to be taken account of during operation than is the case with other activities. This complexity derives from the fact that at basic transformation points major chemical or physical changes in the product occur. For example, different materials may have to be brought together to be synthesised; a complex sequence of activities, observations and measurements may have to be followed to carry out the conversion activity successfully; many different variables, such as temperature, pressure, chemical content, time, etc may have to be taken into account; and often several people have to collaborate closely to get the job done, thus necessitating a complex network of roles and social interactions for all concerned. Storage, transport and ancillary operations rarely share these qualities to the same degree.

Secondly, and as a consequence of the high complexity of factors surrounding a basic transformation, its execution normally involves a correspondingly wide range of decisions. To a greater degree than any other activity in the process, decisions have to be made as to which factor has to be combined with which other factor, in which way, at what time, in which quantities and under what conditions. In the steelmelting process for example, variations in the quality of scrap, in the hot metal, in the state of the furnace lining, and in a host of other variables, have to be taken into account in order to effect the transformation successfully.

Thirdly, basic transformations tend to make greater demands than other parts of the production process on operatives' analytical or numerical skills, on their reasoning ability, and on their capacity for mental attention. These <u>mental demands</u> may also be supplemented by <u>physical demands</u>, such as the need for dexterity and co-ordination of senses with limbs, in so far as the major conversion process calls for manual manipulation by the operater of the materials as they are being processed.

A further characteristic of basic transformation points is that, because the material being processed undergoes a profound change at these points, the process, besides being complex, is also usually <u>irreversible</u>. Any mistake made can therefore have serious consequences, both in terms of wastage of material and in terms of equipment costs if damage to the machinery is involved, and, once again, this is seldom a feature of other parts of the process. Closely related to this is the fact that the possibility of product variance is normally the most critical and most concentrated at basic transformation points.

Again, the nearer to basic transformation, the faster, more frequent, more specific and more accurate the <u>information flow</u> has to be in order that the system can function efficiently. This is because the number and complexity of the variables involved is greater than at points in the process where minor changes are occurring. Similarly, the machinery and equipment involved in the transformation is usually more highly complex than elsewhere and so require more detailed observation and attention for their operation. Even in fully automated sequences, the complexity and number of dials in the central control cabin testifies to the high informational needs of basic transformation points compared with others in the process.

At the other extreme, storage activities demand comparatively little information. As long as one knows on which shelf the object is lying, what the level of stock is and when the next item is required, the storage system will run itself without too much trouble. In this case there is no need to know all about the internal properties of the product since these do not change. All one does need to know is the identity of the product, where it is kept and where it should go next - in other words only those things which relate to those aspects which 'change' in storage and transportation activities, namely time and place.

Finally, basic transformations are intrinsically capable of providing greater motivation(1) for the operatives involved than other parts of the process. We say 'intrinsically' because technology may act as an intervening variable between the operator and the process, and hence reduce his motivational opportunities. Here, however, we are making the assumption that the technology does not significantly inhibit the relationship of the man to the material being processed.

---

(1) We are making the conventional assumptions about causes of motivation eg building skills, variety, challenge, feedback, discretion, growth and learning, etc into the job.

In the first place the high complexity of factors in basic transformations serves to make the operatives' role an interesting and motivating one because the factors in themselves, and their relationship to other factors, have to be <u>intellectually understood</u> by the operatives concerned if efficient production is to be achieved. This requires a knowledge of both the machinery and the materials involved, and also an operational knowledge of such things as product variations, job sequence variations and working procedures; and, because of the complexity of factors involved, the knowledge required at transformation points will normally exceed that which is needed to understand other simpler parts of the process. In addition, therefore, transformation points also offer a greater <u>opportunity for learning</u> and this too has considerable motivational advantages.

Secondly, the wide range of decision-taking inherent in the process at these points entails from the operater a correspondingly wide <u>exercise of judgment and discretion</u> as the process unfolds, and the opportunity to exercise such discretion again has considerable motivational advantages.

In addition, the need for numeric and other mental skills has already been mentioned, as has that of physical dexterity. Again, the opportunity to use skills and abilities are powerful sources of motivation.

Fourthly, the irreversibility of the process at basic transformation points imposes a special <u>responsibility</u> on the operatives involved to perform their task efficiently, and this too can have considerable motivational value.

Fifthly, the centrality of basic transformations from an information point of view means that operatives involved are at the centres of communications, and consequently, are in a <u>socially advantageous</u> position. This again can be motivationally beneficial.

And finally, as the basic transformation is the major change point in the process it makes the most visible and the most significant contribution to the attainment of the overall objective of the production system of which it is a part. It is therefore central in that it gives <u>meaning and reference to surrounding activities</u>, which are all either preparatory or ancillary to it. This again is motivating.

Some of these motivational characteristics will, of course, be found at other points in the production process also, but it is only at basic transformation points that they all coincide and reinforce each other. For example, it may be that a persistent bottleneck somewhere in the process makes another activity the central one for a while, in that surrounding activities have to be scheduled around it. However, in this case, there is no intrinsic reason why operatives should find this other activity interesting to perform <u>per se</u>, or why it should be particularly identifiable with the main production objective, except in a negative sense. Or else it may be that another activity, e.g. driving a crane, involves a fairly extensive use of manual skills, but here again, the activity is neither necessarily socially central, nor is there likely to be much discretion involved.

For all the reasons given so far, it may be said that basic transformation points are the central points in the production process from a motivational, technical, informational and process point of view. Other activities exist to serve them and to support them. They cluster around them and are shaped by them. Moreover, as a consequence of this centrality, it can be shown that the nearer another activity is to the basic transformation, the more technically tightly-knit it becomes with it, and the greater the <u>causal</u> interdependence is between it and the basic transformation. In fact, in every production process there are normally as many 'clusters' of activities as there are basic transformations. Conversely, storage activities, which are the lowest points in the 'change hierarchy', usually mark the boundaries between production systems, and serve as buffers between adjacent production sequences. They are buffers because whenever conflicting requirements for material arise as between adjacent production systems, one can simply add to or subtract from stock until equilibrium is again reached.

Production systems cluster around basic transformation points for two reasons. First, the basic transformation represents the central _purpose_ of a production system. One prepares the iron in order to turn it into steel, and one transports the clay, mixes it, etc, in order to make a pot. Melting and firing are central since they answer the question 'why' the other activities exist; they exist _for_ the basic transformations, to support them, complete them, help them to be realised. But second, the product undergoes the maximum change at the basic transformation point and so _technically_ the input and output activities surrounding it are dependent upon it and have to be tightly integrated with it to facilitate the change. Technically speaking, that is, the basic transformation may be said to _cause_ these surrounding activities to occur. This means that it has a certain power to force its character on other activities in the process. The manner and degree to which it can do this determines the strength of the causal link. Thus an activity can be causally dependent upon a basic transformation because it _always_ precedes or follows it; or because it precedes or follows it _immediately_; or because the way the basic transformation is performed _determines the shape or manner_ on which the other activity is performed. The more intense and numerous these dependencies are in a given situation, the tighter is the causal connection between the two activities.

This 'binding together' of activities surrounding basic transformation points means that in modern industry the practice of management is **not** primarily concerned with supervising individual units of machinery, but with operating _systems_ of production. Indeed the two requirements of a common purpose and technical inter-connectedness are precisely what turn a random group of machines into an organised system. One might say that a modern factory is, in this respect, like the human body with its single heart and bloodstream integrating the function of the limbs. Where the causal connections are perfectly formed the production system will function effectively, where they are not there will be bottlenecks, imbalances and poor machine utilisation.

This is why, over the years, engineers have increasingly attempted to integrate factories into production systems centred around basic transformations. Process routes have been designed so that one machine feeds the next with part-processed materials at the right temperature, in the right state at the right rate in order that the major conversion can take place without fault. They have engineered their products so that one part fits exactly into the whole in the right sequence and position. Much of what goes by the name of automation is an attempt to guarantee the infallibility of this synchronisation still further: by transferring into the technology itself the control of the process so that all activities surrounding basic transformations can be planned and controlled smoothly, without disruptive intervention by outside forces.

The _technical_ notion of a production system defined as the cluster of activities centred around a basic transformation can usefully be linked to the _social_ concept of the primary workgroup.

Small group research has amply demonstrated that participation in decision-taking, success in problem-solving, team productivity, friendship and cohesion, mutual help and support and other motivational and efficiency advantages are highest among small numbers of people and deteriorate markedly when work groups exceed primary workgroup size (say 15-20 people). As industrial work normally requires collaboration, it therefore makes sense to create an organisational structure in which the basic unit is the primary workgroup.

This said, the question arises as to how to relate primary workgroups to production systems. Three possibilities can be envisaged:

(1) no particular relationship

(2) several workgroups in one production system

(3) one workgroup to one production system

In our view there are significant motivational and efficiency advantages in designing work so as to create an organisational and technical structure in which one primary workgroup is made responsible for operating one production system centred around a basic transformation (as in (3) above).

First, as we have seen, each production system consists of an interrelated cluster of machines, and because of this it is also a tightly-knit information system. The technology - and information - is greatest around the basic transformation and least at storage points. Equally, however, from a human point of view, communications, co-ordination and control tend to be better within primary workgroups than between groups which are separated by organisational boundaries. Hence if one workgroup performs activities in one production system centred around a basic transformation there will be a match between the facility for communications and the need for information - high at the centre and low at the boundaries. In any other configuration (as in (1) or (2) above) the match will be distorted and delays and bottlenecks due to 'communications failures' will be more likely.

The second reason for wishing to match one workgroup to one production system is that this is the only way in which all group members can have a chance to be directly associated with a basic transformation, and hence have access to activities of high motivational importance. Such groups, taken each as a whole, would also be able to operate relatively autonomously since they would be technically separated from adjacent groups by storage activities. The greater resulting psychological 'ownership' is also motivating.

Finally, a match between a workgroup and a production system has important benefits for labour productivity in so far as it provides the ideal conditions for flexible working. From a social point of view, as has already been mentioned, informal flexibility is much easier to achieve within a workgroup than between groups. But from a technical standpoint also, a production system consisting of a process-based longitudinal slice of work centred on a basic transformation provides a patterned variety of complementary jobs which maximises the opportunity for flexibility in relation to the work being done.

Unfortunately it must be admitted that in much of modern industry today, a 'match' between people and technology such as we have been advocating does not exist. Often the scale and labour intensity of production is such that instead of one workgroup one finds several groups immersed in executing a single production system. The motivational and efficiency disadvantages of this state of affairs are obvious, and one can only hope that plant designers will in future pay more attention to matching the two before inefficiency and poor labour relations are literally built into the plants under construction. What is needed, in short, is a mix between labour and capital which both maximises the technical economies of scale and maintains the one group-one production system identity.

Finally, in order to ensure that each group member <u>actually</u> is able to participate meaningfully and productively in the group's task, it will also be necessary to examine the process-technology-operator relationship closely to obtain the best allocation of function between men and machines. A useful starting hypothesis for such an examination might be that:

(1) Lower-order work activities such as those involving simple energy, the carrying out of routine instructions and actions and self-measuring and adjusting are best performed by machines; while higher-order activities such as those involving originality, creativity, judgment, higher reasoning and learning are best performed by man; and

(2) the same higher-order activities best performed by man are those which are also the most motivating to him.

Whether these hypotheses can be verified in practice, and if so, whether they can be combined into the group task structure suggested above to provide the foundations for future plant design, remains to be seen. If they can, the basis for a new industrial order compatible with human needs could be laid.

(C) CHRISTIAN SCHUMACHER, JULY 1977

JOBS AND VDU'S, A MODEL APPROACH

S. Scholtens and A.J. Keja

Hoogovens Estel

INTRODUCTION

Very seldom in papers on VDU's reference is made to the different possibilities in using the device.
Based on practice 4 models of interaction can be recognized. Each model has its own characteristics and makes specific demands upon the user. So if we are going to discuss "Jobs and VDU'S" it will be useful to describe the models.
We must be well aware of the fact that these models are simplifications and also that more models can be developed by combining part of the one and part of the other. Nevertheless 4 basic models to be presented are worth concentrating on.

MODELS

1.1 Interaction model # 1

In the flow diagram we recognize two parts as drawn up on the left hand side - man - and on the right hand side - information -. It becomes clear from the diagram that the main purpose of this setup is information display. Information, which the designer expects the user will need to do his job.

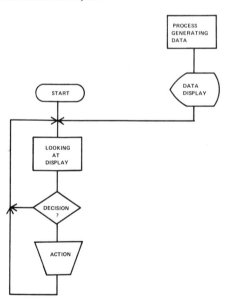

INTERACTION MODEL # 1

To present information in this way - by using a VDU - is decided upon because the data can easily be distributed and kept updated. Anyhow if we compare the possibilities with the classic methods in sending around by post-service sheets of data generated some day before their actual use, this is an improvement.

However, the disadvantage of this presentation of information compared with the classic paper system is that you cannot take the data with you unless you generate, in one way or another, a hard copy for instance using paper and pencil.
Characteristic for this model is that the user has to wait till the data he needs, appear on the screen, requiring a permanent attention drawn to the displayed data. This can be recognized if we think of the displays of departure times - with further instructions for instance BOARDING NOW and gate numbers - in modern air field stations. Particularly when there is some disturbance in traffic a lot of people are glued to this display!
However, the experienced traveller has built up his own model of expectance concerning changes in data. He finds clues that inform him, or preinform him, his attention must go at a certain moment to the said notice board.
If the situation differs from the ones he based his experimental model on, he will likely miss essential information. In the example of the air field there is, a back-up system. The public address system is then crying out loudly: "Will Mr. Experienced Traveller inmediately go to gate No. X".
Essential in the example is that there is available a mode with which attention can be drawn from a user to the information.

Observing the model in practice we found that a number of times information was missed, either recognized too late or missed altogether. For example chages in data were not recognized because these changes were not expected. Remedy: increase the recognizability by either colour labelling, reverse background or flashing. The period over which this additional label has to stand up, has to be limited.
But how to limit this additional label? Of course there is a very practical criterion: the additional label may not appear when no longer special attention is asked for this labelled data. This results in, the system once having labelled an item, to check at adequate intervals if the label is still valid.

If the performance of the main task, for which this information is displayed, depends on this information, this interaction model may lead to a stress situation for the user. Perhaps he will adapt to this situation, the consequence is then losses for the system.
Anyhow we can state that the information is there as the designer intended tot present, but not recognizing the user as a human being, the performance of this man/machine system is poor. It may be cheap but at what price.

1.2 Comments on model # 1

To draw the users a attention to changes of information on the display, it is possible to use the "reverse background"-facility that many displays have got.
The software consequences are that in the "data-space" in front of every item (or field) that has to be displayed, room has to be allowed for extra information concerning "reverse" or "normal" mode of displaying the item that has changed recently. The extra room for this information may take up to 10% of the total space for all data. Using colour causes no real differences from a software point of view, neither does flashing. From a hardware point of view it

can be stated that colour-displays are much more expensive than black-and-white ones, so one has to ask oneself the question whether the use of colour will really be an improvement compared to reverse background or flashing of the critical items.

Another difficulty in this model # 1 is that the operator has no way of telling the computer that he has seen the critical item, so the extra information by way of colour or one of the other possibilities can be removed. The only way this can be done, is on a time-base. After a certain period of time the computer simply decides to "set the item back to normal". Hopefully the operator has seen the critical item during this period of time, but there is no way to be sure of this.

When we solve this limitation on a fixed time base, then we have to decide on the duration. In analysing the main task, the time lap between two fixations of the man on the screen can be found and for instance the time base could be chosen 3 to 4 times this period. Of course there are disadvantages in using this derective. For example when there is a large deviation between the 5th and 95th percentile of the fixation intervals, then under certain conditions the label is annoyingly long on if we choose for 4 times the 95 percentile value.

Another way of limiting the label is to say just one label, perhaps the latest one or the most important, will appear on the screen. In this case we accept loss of information. Anyhow the conclusion can be made that solving this problem is only possible if we ana' lyse thoroughly the job of the man and the way he has to make use of the data displayed. Only then we can compromise on what we really want to transmit and what really can be absorbed by the user. This method is only valid, if not too many changes occur. Still we cannot be sure whether the information is passed to the user since there is no way in the model to inform the system that the information is understood and accepted.

1.3 Summarizing model # 1

The interaction model # 1 consists out of data display only. The possibility to update the displayed information in an easy way on a short time base is an advantage.

Since there is no direct feed back into the system, the application has to be in such a way, that the use of data is incorporated in the job.

Attention has to be given to the way changes are displayed a recognizable way.

2.1 Interaction model # 2

In this flow diagram we also recognize two parts. The difference compared with model # 1 is that there is an influence from the user on the data displayed.

The essential point here is that the user can call for data any moment he thinks he needs data. In this case the model stands for a search system.

The system will be used the moment the user thinks that the system can give him that information that he needs to perform his task.

If the search, the system must perform, is simply a selection of one out of a limited number of possibilities that are known to the user and to the system, this will greatly simplify the software that is necessary.

It is obvious that an outside impulse has to activate the user to consult the data bank. It is of utmost importance for the success of actions resulting from this urge, an answer in one way or another will be generated. Or, if the input is insufficient determinative, a prompt is presented on the basis of which a more sophisticated search for the answer can be carried out.

A valid example in this area is the telephone directory. Imagine someone is calling the enquiry service asking the extension number of a Mr. Smith. Imputting the bare family - name Smith into the system, would generate a large number of Smithes, perhaps more than adequately could be selected from, in order to give the right answer to the caller.

It will be quite clear that an additional "SORTING SYSTEM" is

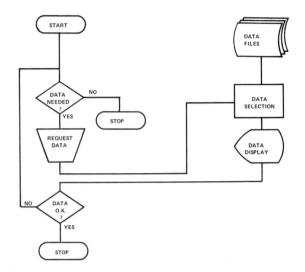

INTERACTION MODEL # 2

needed to come to the Mr. Smith, whose extension number is asked for.

If at this moment "UNSUFFICIENT DATA" is appearing on the screen this will not help the user much further in solving his problem.

And here the problems arise. What additional data are available for selecting the Mr. Smith. His Christian names, the area he lives in, the house number, the street name or perhaps the family name of his wife?

There are many possibilities on the basis of which you can come closer to the right information. We cannot expect that the caller can generate this more specific information in a fixed order. So the only thing the telephone operator can do, is ask for more specific clues to come to the answer.

Depending on the clue, the system has to react with either the right answer or a question for more specific input data, or perhaps gives some suggestions. Depending on the service you want to offer, the scheme will be more or less complex. In the case of the inquiry service of a telephone company, it is clearly understood that they aim at a high level of service.

Even quite distorted basic information should lead to adequate answers. In terms of programming this results in multi access systems. It will be understood that the moment the answer can be generated no further input is needed. In other words all the possibilities will hardly ever be used.

And now the problem this type of communication with data banks puts to the system designer. Of course we can accept that the user can be trained in such a way that unco-ordinated requests on the system will not be expressed.

We must be aware of the fact that analysing future needs - in this case what requests for data will be formulated by the applicant - is a very difficult job. This analysis cannot be based on the assumption of the not yet existing system. Acceptance of a system, however, largely depends on the performance of such a system, in other words to what extent the program is adaptive to newly generated demands to this system. What aids are available accessing the data bank?

A way to solve this problem is to give the user full liberty in leafing through these data pages. The disadvantage of this method is that the amount of data is so large that this leads to no result at all.

As far as changes in the data occur, there is now a possibility for the man to give an accept. In this case there is a better garantee the user has recognized the change, then there was in the case of the # 1 model.

2.2 Comments on model # 2

On the matter of the multi-access-system for the telephone company inquiry-service the following can be stated.

Suppose that the organization is such, that a file exists for every familyname. Within this file the family-names are arranged by city and by streetname.

Example

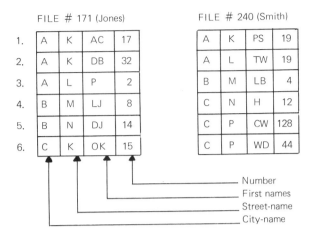

A lot of problems can be solved in an easy way, if this organization is used, even if the city-name is unknown (or any other item; as long as the last name is known).

Question: can you give me the telephone-number of a man named D.B. Jones living in a street named K (I do not know what city)?

The software has to scan the Jones file for street-name K and first names D.B. and after that, it can display the name(s) that it has found, even if 1% of all people have the last name Jones and there are 2.10^6 subscribers it will be a matter of scanning 20,000 records which is an operation that can be performed in say 15 - 20 seconds. However, as soon as the last name of D.B. Jones is unknown and the system has to look for Mr. D.B. X living in a street named K in a city named A, this will be quite a different matter. The scanning necessary to perform this action will take say 1500 - 2000 seconds, which is unacceptable to any operator (and to the system).

A way to solve this problem is to maintain a second (and perhaps a third and so on, but let us limit ourselves to a second) "data base" with relational information of perhaps the following format:

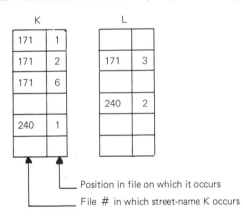

This "data base" would be as large as say 10^7 words of memory in the case of 2.10^6 subscribers, not to speak of the software consequences of updating and removing entries from two or more related "data bases".

2.3 Summarizing model # 2

The interaction model # 2 aims at selecting specific data out of a file. Depending on the complexity of the data required and on the complexity of the information on the base of which the data are going to be searched for, it will be necessary to have different access methods. In complex files help programs will be needed. The way data will be presented, depends on the way the search task is carried out. An analysis of how users do their specific job has to be carried out, before setting up the files and decide on the display lay-out.

3.1 Interaction model # 3

This third model differs from # 2 in having a possibility to change the data. As goes for all the models, this one is simplified. Here the task of the user is to update a data file. The new data to be brought in can be presented to him by way of written material. From this written material he selects a workable set. Via the display he selects the file to be updated.
And then the "real" work starts. A lot of what was said in models # 1 and # 2 is also applicable for this model # 3.

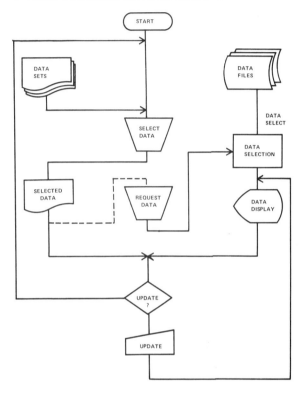

INTERACTION MODEL # 3

The new element here is the matching of data sets presented in one way with machine generated data.
Decisive for the machine program has to be the way these new data are presented. Can these easily be grouped in such a way that fast input is possible? Not too many steps between two inputs? Are the data to be brought in of the same type on paper as they are on the display or has the operator to transform these data?
For example the data to be brought in are of the type of guilders per square meter while the file only accepts prices per kilogram. In this case it would be worthwhile to have this transformation done during input and not to give this additional load to the man. This is the more important because the documents are generated outside the computer area and thus there is a certain freedom in the way these are filled out.
Certain rules and guidelines must be there, true, but do we have to force everybody into the monocultus of the computer? It will be better to program the computer in such a way that it is a help and not a restriction in our daily work. Another example in this field is the input of words. In our company we use coke for the production of iron. In a stock control system we had some difficulties during start-up. What was the case?
The people working on the stock control in the manual way used for coke the following three words: "coke", "kooks" and "kks". There had never been any difficulty in these words until the workers got via a terminal direct access to the files. The computer only accepts the word "cok".
From program efficiency point of view this can be understood, but is this restriction really necessary? I do think that for full

time experienced users this limitation can easily be coped with. But the development is that more and more people will use terminals not as a full time job.

We can compare the new situation with communication by telephone. In our model the communication is not between people but between people and a machine via an apparatus that is much more difficult to handle than the telephone.

The assumption that more people will make use of terminals in their work results in additional criteria for the program and the limitation resulting from this program for the user.

We alle know that everyone has his own way of expressing or perhaps better of thinking. And we also know that a good communication only then is possible, when the communicating men are "tuned" to another.

There where VDU work is the communication between user and machine (read: programmer) we should like to have some "tuning facilities" in this machine: "adaptive to the specific user".

3.2 Comments on model # 3

The problem, that every operator wants his own way of "talking to the computer", can be solved in a number of ways, the easiest of which is to decide in advance which operator wants what. It will be possible to develop different screen lay-outs with the same contents of information, but with different ways of entering new data into the computer.

However, the software consequences are enormous. For every screen lay-out that is wanted, it is unavoidable that say 1 - 2 K words describing the layout is reserved on mass-memory and the programs that have to be run whenever some item is entered or altered, are different. So for every screeen layout an average of 6 - 8 K words of programming is needed.

If one considers a 4-shift operators team this could mean that instead of one screen layóut, 4 (or 5) different ones have to be programmed, so the costs would be very high.
This is not equally true in the case of entering different codes, having the same meaning.
It is of course possible to program the computer in such a way, that it gives the unexperienced operator some help when he makes an error (from the programmer's point of view).

One of the ways of doing this, is to display all possible entries if the operator did not key in one of the allowed entries. (Simply display the buffer that the program has got, to be able to decide whether the entry was allowed or not).
A next level of sophistication could be to allow the operator to enter new codes into this buffer by stating for example: COK = KKS; in this way telling the computer dynamically to accept both answers from now on. Although this looks much more simple than programming all possibilities in advance, there is more to it. Software has to be written to make sure that after a "system breakdown" the process of teaching the computer what answers to accept, has not to be started all over again. The consequences of this way of treating the operator as a human being instead as a preprogrammed keyboard puncher are smaller than if one gave every operator full freedom to choose his own way of communicating with the computer.

3.3 Summarizing model # 3

The interaction model # 3 is mainly used for data input activities. From an economical point of view the way data is inputed will be according to strickt rules, thus limiting the user in making his own dicisions on how to carry out his task.
Combined with the relative fast response of the system the period over which the user can carry out his task will be limited. Combining tasks of a different nature with the model # 3 is advisable.

4.1 Interaction model # 4

Essential in this model is that it generates new data, the validty of which the user has to decide upon.
The way of working is that the user takes during the process a great number of decisions in each step in the program. The machine is very fast indeed, the moment the decision is made, a new decision is asked for.

For example making a complex stress calculation in a construction. Machines are working at such speeds that one man can perform the work of 70 manyear manual calculating in let us say a few days. But the decisions have to be taken. And here we are confronted with the mental capacity of the user. How long a period can he continuously work in this interaction model and if this is limited, this can be initiated by the program.

We just want to say that complex programs have to be built up in such a way that they can be broken off at any convenient moment from the point of view of the operator. It cannot be that this results in a wate of a days work.

4.2 Commentes on interaction model # 4

The wish to break off a program in progress at any convenient pont and of course to come into the program after some time without losses, forces the programmer to breake up his program in a relative large number of short logic-blocks. To organize these blocks he needs a number of controlling programs. Moreover additional store is required for data and recording the steps carried out. So this leads to a number off steps adding up to a slow down in performance. The question is then how slow is acceptable. Do we take into account that users with different capacities will use the program and that fatigue effects will change the capacity of any user, then this complicates the answer. Anyhow the programmer is forced to make his program less efficient then would be possible. This also goes for the way data will be displayed and set up of the conversation.

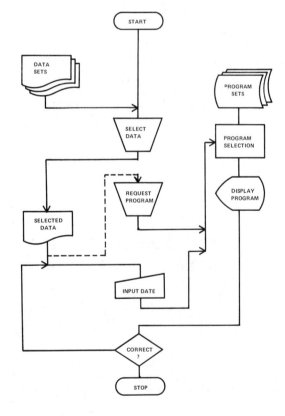

INTERACTION MODEL # 4

4.3 Summarizing interaction model # 4

The interaction model # 4 stands, for calculating activities by means of computers. The operations to be programmed, displayed and activated mainly depend on the capacities of the user.

5. Comparison between the models

A relative comparison over the models # 1 to # 4 demonstrates the difference between the models. The models are compared on the basis of:

- data choice : is there a possibility for the user to make his own decision in what data are displayed;
- data recognition : is there a garantee the user will recognize the presented data and data changes;
- communication patterns : is the communication with the system complex or simple;
- user's decisions : can the user decide on when and in what order to carry out his task;
- lay-out displayed data : is the lay-out of the displayed data important for a gooed performance;
- memory capacity : in a technical way, is a large or small memory capacity needed;
- program complexity : how complex the program has to be;
- back-up need : necessity of having a back-up.

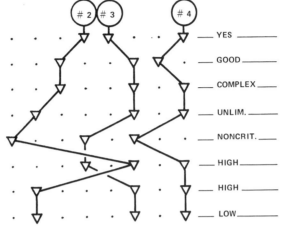

6.1 Use of interaction models

From the examples given in the description of the interaction models can be found that each model has its own set of conditions for program and lay-out of the displayed data but also for the lay-out of the workstation. Moreover the use of VDU's according to any of the models, influences the job structure.

Now real systems consist in general out of a number of workstations. It will be clear that a model # 1 station always has to have a model # 3 input station. Also other combinations can be set up. Examples are given in the diagrams.

If we take into account the demands that each model sets, for example for the data lay-out, this leads to the necessity of conversion.

If, and that nearly always is the case, more then one user is influencing the data (multi interaction diagram # 3 # 3) special attention has to be given to the interference of work of the one and the other.

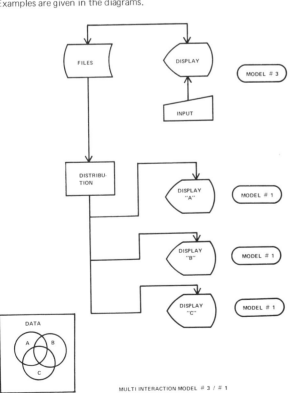

MULTI INTERACTION MODEL # 3 / # 1

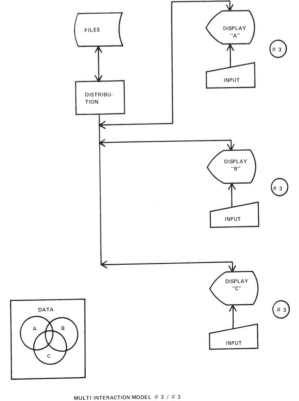

MULTI INTERACTION MODEL # 3 / # 3

Very important is then that the data changes, with consequences for the task of one of the users, are clearly signalled to this user. It may even be necessary to inform an user that the frame he has selected at that moment is selected and worked upon by someone else. Help programs will be needed to inform a user were changes by others influence his decision making. To give the user the possibility to send additional information, this will mainly be done in off-normal situations, it can be worthwhile to reserve on the screen a field where any message can be written.

The advantage of systems with VDU's (but also comparable devices) is that data is transmitted by line. This resulting in the effect that workstations can easily be separated over longer distances. We must be aware that this may lead to isolation. Remedy is to combine jobs.

6.2 Comments on multi-interactions

From a software point of view the easiest way of treating interaction is to forbid it. This could mean that two operators A and B, both responsible for controlling part of a process can only use the system one at a time, whenever they want to use it for activities that could cause inference.
The disadvantages are clear. Whenever A "forgets" to end his activities, B must inform him (by telephone or so) thtat he wants to use the system. As soon as A alters some of the data, he is responsible for it and the system does not inform B that this has happened; a chain of errors and mistakes will result.
This is far from solving the problem of interference.

An easy way of informing B of what A has done, is to update B's screen every x minutes; x is depending on the time-scale of the process that A and B are controlling. In this way B can be sure that he looks at the right information every x minutes, but the problem of informing B that A has made an alteration in some item(s) still exists. B has to scan the screen for these alteration, which he can easily miss.

The best way of solving interactions is therefore to ask B's attention the very moment A has changed an item on his screen that also appears on B's screen.
B can push a button on the keyboard confirming that he has seen the altered item or take a comparative action.
The software problem of solving interactions like this, is that the computer must know whether the item that has recently been changed, appears on other video screens than the one, it has been changed from.
If operator B has 4 possible screen lay-outs he can select and on 2 of these layouts the item in question appears, the software must remember on what screen layouts the item can appear and after that a search must be made, whether these screen lay-outs are actually displayed at this moment and where.
The next thing to do, is to update the appropriate item at the same time informing the operator that the item has been changed. It has been found that this way of solving interaction is really necessary in some cases, where fast decisions have to be made by two or more operators.

6.3 Summarizing multi-interaction models.

A VDU station is never an independent piece of equipment. As part of a system it has to fit in the system. However, a VDU workstation belonging to a system will have its own considerations concerning communication scheme, information lay-out and input devices. These considerations will result from a analysis of the task to be performed.

Environmental influences may limit the task performance and thus may influence the considerations as well as workspace lay-out.

Whatever task conditions may exist, those will determine the interaction models to be chosen for a system.
Within a system different models can be used.

REFERENCES

Birren, J.E. (Ed.) **Handbook of aging and the individual,** Chicago: University of Chicago Press, 1959.

Welford, A.T. Changes in the speed of performance with age and their industrial significance, **Ergonomics,** 1962, 5, **1,** 139 - 145.

Ballantine, R.M., Penniall, T.H.
Information Systems design - display aspects
CEL-BSC London 1974.

Work Stations with data terminals,
Stansaabs Elektronik AB
Järfälla Sweden 1974.

Haider, M. Slezak, H. Arbeitsbeanspruchung un Augenbelastung an Bildschirmgeräten.
Gewerkschaft der Privatangestellten.
Wien 1975.

Bell, R.A.
Principles of Cathode-ray Tubes, Phosphors and High-Speed Oscillography.
(Selection III) Application Note 115, Hewlett Packard
Colorado Springs (USA) 1970.

Emery F., Thorsrud E.
Democracy at work,
The report of the Norwegian industrial democracy program, Australian National University, 1975.

Broadbent, D.E.
Vigilance, 1964
In: Holding, D.H., **Experimental psychology in industry,** Penguin, 1969, 167 - 179.

Buckner, D.N. and McGrath, J.J.,
Vigilance: A Symposium Mc Graw-Hill Book Company Inc., N.Y. 1963.

Davies, D.R., Tune, G.S.
Human vigilance performance, Staples Press, 1970.

Tickner, A.H., Poulton, E.C., Copeman, A.K. and Simmonds, D.C.V.
Monitoring 16 television screens showing little movement, **Ergonomics,** 1972, 15, **3,** 279 - 291.

Hick, W.E.
On the rate of gain of information, **Journal of Experimental Psychology,** 1952, 4, 11 - 26.

Welford, A.T.
What is the basis of choice reaction-time? **Ergonomics,** 1971, 14, **6,** 679 - 693.

Stewart, T.F.M.
Display system design for improved operator organisation,
IEE Conference Publication Number 150, 82 - 85.

Umbers, J.G., **CRT/TV Displays in the control of process plant: a review of applications and human factors design criteria,** Warren Spring Laboratory (UK) (1976) (ISBN 0 856240 91 5).

SYSTEM DEVELOPMENT AND HUMAN CONSEQUENCES IN THE STEEL INDUSTRY

K.S. Bibby, G.N. Brander and T.H. Penniall

Human Factors Department, British Steel Corporation, 140 Battersea Park Road, London, England

INTRODUCTION

This paper represents one picture of current developments in automation schemes and their significance upon the progress of a Human Factors Department pursuing "job design" activities within the British Steel Corporation.

We present our own assessment of our progress, as a basis from which to comment on the nature of both the technical and organisational features affecting the work of an applied research group; and with a view to stimulating discussion during the Workshop, concerning the organisational determinants of making such a group a vehicle for successful change within an organisation.

The group evolved some twenty years ago as a result of a synergic reaction between industrial developments and people who were becoming more aware of the human factor problems. From the narrow ergonomic considerations of the various aspects of workplace design during the early days, the group has moved towards a broad application of the "job design" concept. The group now has almost two hundred man years of effort behind it of which twenty percent, at most, has been in the nature of an 'in-house' consultancy service responding to external requests from the industry. The major part of the effort has been expended in developing a knowledge base which has led to improvements in the efficiency of our service to the industry. During the last five years, as a team of social and behavioural scientists, we have endeavoured to reconcile our research activities with our industrial base through performing case studies and investigations both for the immediate benefit of the site involved and, more importantly, in order to permit the formulation of more generalised recommendations and guidelines for the whole industry.

Developments in process technology and evidence of the potential advantages to be gained from computerisation are both areas which have been exploited within the steel industry and, as a result, have created profound changes in the mechanics of production. Illustrations of some of the technological developments which have been implemented serve to create a perspective against which the activities of the Human Factors Department need to be considered; insofar as the introduction of new control capability has had far-reaching effects within the organisation. It is our reaction to these changes as an interested unit within the organisation which forms the basis of this paper.

THE INDUSTRY - A BACKGROUND

The industry in which we work comprises three fundamental parts, each with its own objectives and constraints.

1. Ironmaking takes place in blast furnaces. There are usually three furnaces in operation, each of which is tapped on something like a three hourly cycle, though this is variable, and largely unpredictable. There is a considerable lag (possibly as much as 24 hours) in the process between corrective actions being taken, and the hot metal analysis reaching a new steady state.

2. Steelmaking is largely by the BOS route (80% of BSC's steelmaking capacity) where a cycle time of about 40 minutes is common. This process occurs in a three or four vessel shop, with one vessel off for relining etc. There may be a buffer stock of iron held in mixers. The casting of steel into ingots also takes place here.

3. Rolling and finishing are buffered from steelmaking by the soaking pits, where ingots are heated to rolling temperatures.

Blast furnaces operate most stably at high throughout and there is no option but to tap a furnace when it is ready. Steelmaking plants prefer to make steel and get it out of the shop as quickly as possible, due to the enormous complications which may result if delays occur. The steelmakers, therefore, tend to exercise an arbitrary control over the order in which they choose to make a required list of casts, and sometimes make qualities for which there is no order, to save time when they face operational difficulties. In the mills, the ideal rolling programme is one with long runs of particular qualities and sizes, though the very varied nature of the BSC's order book makes this a rare occurrence.

THE INDUSTRY - THE CHANGING TECHNOLOGY

A typical example of the changing emphasis within the industry brought about by significant developments in process technology has been the rapid replacement of the open hearth (OH) steelmaking process by basic oxygen steelmaking (BOS), or the Linz-Donawitz process. In the UK, from less than 1% of steelmaking capacity by the BOS route in 1960, when OH accounted for 86%, the proportions have changed to 32% and 46% in 1970, and to 75% and 5% respectively in 1976. Production cycle time of one furnace, typically around 350 tonnes, has decreased from about ten hours (OH) to forty-five minutes (BOS). More liquid steel is passing into an alternative process to ingot casting, i.e. to continuous casting plant, for which target steel parameters are more critical. In consequence process control in the BOS process assumes increasing importance. From static process models, the trend in development is towards dynamic modelling for better control to be exercised throughout the course of the steelmaking cycle.

In the field of automation techniques it is only thirty five years since servomechanism theory became widely accepted in process control schemes which, previously, had been based largely upon pneumatic instrumentation. The early 1950s were marked by a strong belief in centralised control, with the consequent emergence of vast instrument and control panels. Since these early days, there have been considerable technological developments :

- the early 1960s were characterised by the advent of electronic instrumentation although there was still much reliance upon pneumatic instrumentation

- the last 1960s showed an increasing acceptance of electronic instruments primarily due to the reliability of the new semi-conductor devices

- the early 1970s were marked by the transition to integrated circuit technology.

At this stage, however, it was becoming apparent firstly that primary process transducers were considerably under-developed by comparison with control and instrumentation devices, and secondly that control philosophies were in need of reflecting advances in control theory.

In recent years energy conservation has become another important factor in production. There is now little scope for reducing energy consumption by individual processes and the trend is towards energy monitoring and control of all the inter-related plants on a works for further economies to be viable.

As illustrated by the example of energy control, automation is now beginning to expand beyond process control, and through linking processes together into decision networks, plant control now becomes a real possibility. On the other hand, the past few years have seen a trend away from large computer installations performing many functions, towards distributed mini-computer systems, and also new methods of applying economic control to process plant this being primarily due to external economic pressures). Technical rationalisation to meet the demands imposed by the contraction of domestic and world markets, coupled with a rapid growth in the sophistication of the market, have resulted in the development of large, high throughput, complex plants, whose economic breakeven point is, in most cases, at a much higher percentage of designed capacity than ever before. The emphasis on hierarchical control rather than regulatory plant control, the development of instrumentation and control systems based upon digital data transmission, and the recent advent of microprocessor techniques, permit wider flexibility in the design concept of control schemes but this flexibility remains largely unexploited. However these developments have set greater problems for the plant designers as these schemes necessarily now affect a wider range of personnel and functions.

THE "PEOPLE PROBLEM"

We will now proceed to discuss our interpretation of the impact of these innovations in process technology and automation, upon human factor issues, with the implicit suggestion that many industries, although operating above the breakeven point, are operating in a region of diminishing return because of the inadequate design of individual jobs and human systems to release the potential economic benefits of the technical system.

Larger plants increase the physical separation of man and man, worker and manager. In highly automated plants responsibility is devolved away from managers, yet concentrated upon fewer men in the workforce, who are, by virtue of the information at their disposal, better fitted to make most decisions affecting production.

As private citizens in the market place, in the context of increased consumer protection and information, workers have learned to exercise well informed choices. In the work context, information systems make the production situation more transparent (and, where systems are well designed, more easily understood) and hence create another forum for the exercise of personal responsibility.

After a decade in which increasingly sophisticated process control systems have degraded the role, and the skills of the operator, the position has now been reversed, and men are being placed in supervisory roles, monitoring the functions of computer systems with virtually no direct contact with process operations (e.g. in high speed rolling mills)

Often changes are attempted which require the imposition of a "controller", with machine support, who is to take over a set of functions previously performed as a significant part of a senior operator's job, whilst expecting them to accept his evaluation of situations, and to comply with his instructions about how the process should be run. The fact that such controllers are often not expert in the process area often creates problems in their working relationships with the old hands, and these are often further compounded by the definition of these controller jobs as "staff" jobs, with the job holder sometimes selected, and often recruited from outside, by management, instead of coming up through a job progression in the industry. This problem arises partly out of the over restrictive, but historically inevitable union attitudes towards the protection of jobs, which allows seniority to precede merit as a criterion for promotion.

There are many other instances where formerly cohesive work groups have been fragmented and dissipated through the imposition of a 'system' which performs many of their tasks, often more efficiently, but which fails to provide a satisfactory mechanism for knitting together the fabric of social relations within the work which it has disrupted.

THE SITUATION NOW

We can point to piecemeal attempts at improvement. Scheduling systems which, though technically perfect, and offering the information base upon which effective control could be exercised, are, nevertheless, placed in contexts where control over input, and forecasts about likely output requirements, are negligible. These problems are not due to any necessary conditions, but are due to the reluctance of organisations and individuals, geared for many years to process control, and local optimisation, to sacrifice these to broader considerations.

Process control requires knowledge or a black box model of a process; it needs specific process objectives, and the process operator as assistant to act upon emergency states to redefine set points, or to otherwise override the machine function, where the actions required are outside the machines "control region". This whole philosophy aims at local optimisation of specific functions.

In production planning, however, networks of interdependent, contingent events, have to be planned in such a way that the production situation is generally manageable. This may involve overriding, though not violating local criteria and objectives, in the interests of more general plant goals.

The design of socio-technical systems to reconcile these two sets of demands is a destination which, in some cases has not been recognised, which we have certainly not reached, and there is, as yet, little evidence that the two routes described above are converging.

In new plant design, opportunities have existed for management to model the constraints upon their plants, to calculate "least through costs" for their products through the correct selection of raw materials and process routes for particular qualities, and to make in process changes to the product route in the event of plant failures. Given that the steel production process is quintessentially a series of linked, and fairly frequent crises, one would have expected management to have welcomed these opportunities with open arms.

Quite the reverse has been true. In the areas of practical ergonomic advice, where they recognise expertise, and in the area of safety, where they lack expertise but must recognise the pressures of legislation, the way is open; but when invited to examine their conservative production strategies and philosophies, to consider and embrace the full implications of the enhanced flexibility inherent in new systems capability, they draw back, largely, one suspects, out of a typically innate British conservatism.

This raises the question as to where the pressure from change should come from, what changes should be looked for, and how should they be assessed.

WHAT NEXT?

In our time, there are several parties which have an interest in the outcomes, each with their own values, and it is the interplay of these values and the power behind them that dictates the possible and likely outcomes, and it is these values in terms of which outcomes must be assessed.

Technical success, and economic success are pragmatic criteria, which are easily defined and applied against their own contexts. Resulting improvements to the physical environment are likewise open criteria where industry is concerned (except, perhaps in terms of the effects of some of the commodities produced).

Gains in knowledge, favourable changes in social climate at work, or more generally, or more liberal/humanitarian shifts in the general climate of values or opinion are more difficult to define, achieve or assess.

The actors: the piece include employees and unions, in one relationship, the whole paraphernalia of technical and economic research development, control engineers, and social scientists, in our various institutions and organisations, and, the common denominator, which is often forgotten, the whole mass of workers as collectives and individuals.

Asked to define an ordered set of desirable objectives for change in the industrial context, or a set of criteria for assessing recent changes, each group would probably produce very different orderings amongst the criteria. Given that differences of view exist, the sensible resolution of the problem would appear to lie in identifying the stakes each party may have in a venture before projects get actively underway, and defining some positive sum games as the objectives. Recognising that this may not be possible in the case of every exercise, but that changes in some directions may need to occur before the conditions are right for other changes, efforts may be made to ensure that a proper balance of interest is maintained through time.

The role of the applied research organisation in all this is not free from problems. In working to define a way forward in the knowledge of events, techniques and attitudes, it is necessary that applied researchers should preserve a broad view of social and organisational trends, as well as possibilities and constraints inherent in their applications environment. This makes dialogue with senior managers and policy makers essential. On the other hand, direct involvement with the detail of short run applied problems is necessary in order to gain both credentials and allies at all levels. In this context, it is valuable to identify a local change issue which has potential implications for broad policy changes, and which must capture the interest of senior policy makers. For whilst, in general, managers at all levels recognise the pursuit of knowledge as a legitimate concern for researchers, and may well provide facilities to this end, where this framework of knowledge is unfamiliar, or worse still, where it presents them with diagnosis of problems (which they recognise only too well, in general), with inadequate back up for solutions to be found, they seldom make the transition from perception to action, and the contact with the researcher is allowed to lapse. On the other hand, immersion in detailed short run projects to the exclusion of other considerations means that potentially valuable inputs to policy formulation are not being made. Consultants often achieve an audience, access to problem environments and achieve implementation of solutions which are sometimes less than clever, merely by virtue of their public image, claims to general competence, and extremely prestigious costs. Following that example, the right to be heard at senior levels in organisations, and to express competent opinions which take account of organisational constraints and worries, that right, once exercised, becomes self fulfilling.

Humanisation of work, and the harnessing of man and technology in a symbiotic relationship, cannot be achieved without a substantial investment of resources in the pursuit of human and social objectives, and we cannot be said to have succeeded in getting this philosophy accepted until policy statements are backed up with budgetary items on each capital investment programme for such enhancements. The authority to sanction such items, and such a commitment, resides only at the top of every large organisation and government department.

3. NUMERICAL CONTROL, ASSEMBLY AND ROBOTS

WORK ORGANISATION WITH MULTI-PURPOSE ASSEMBLY ROBOTS

Mario Misul

Direzione Metodi di Produzione, c/o Ing. C. Olivetti & C.S.p.A., Ivea, Italy

"Handling of parts and portable tools is quite an important percentage of productive activities and it absorbs a large quantity of manual labor in repetitive and insignificant activities". I think that in this statement of Prof. Umberto Pellegrini found in his preface to the publication "Robotica e automatismi multiscopo per industrie manufatturiere" we recognize one of the principal reason of the ever growing importance of industrial robots.
At least in its initial stage, it is expected that a major application of robots will be in the field of simple and repetitive assembling activities which are tedious and give little possibility of job enrichment to the worker performing them.
Today more than ever, those who are concerned with work organization know that it is necessary to consider both the technical aspects and the social context in which one operates. The worker no longer sees his work as a means of satisfying his primary needs. These, in most cases, have already been satisfied and the worker wants a more meaningful job. He is less inclined to accept jobs with poor content simply against some economic advantage. It is also necessary to consider the higher educational level the new generation of workers has. Young workers are less inclined to accept short, repetitive and simple jobs; they ask for a place where their dignity will not suffer and where their abilities can be expressed. The use of industrial robots can help in relieving man from highly repetitive and meaningless jobs and give him the opportunity to apply himself in a creative manner.
Traditional automation (or rigid automation) has the characteristic of being designed for a specific product. Its applications can be economically convenient when the cost of the machine can be spread over a large production of units and when during the product life no significant changes or variations, that would keep the machine inactive or imply costly adjustments, are foreseen.

The real situation in an organization like Olivetti's is substantially different:
- the market is more and more pretending and continually it asks for more diversified and sophisticated products. It is necessary to respond in due time to new requirements and continuouly provide new competitive approches through the renewal of products.
Consequently products have a shorter life than in the past, thus it is necessary to shorten the start up period;

- the products are therefore carachterized by a great variability in time both for production volumes and models mix;

- the technological evolution further reduces the product life.

All these factors together, plus the increasing labor cost and the need to reduce production costs in order to meet competition on foreign markets where Olivetti exports 75% of its production, force us to develop new organizational solutions.
The first solution adopted by Olivetti to meet this new sociotechnical context is the U.M.I. organization (Unità di Montaggio Integrate = Integrated Assembly Units). The U.M.I. can be synthetically defined as a productive units consisting of a group of workers to whom common objectives of quality and quantity have been assigned and which have been supplied with the necessary resources and information to realise, control and meet the assigned objectives.
Each worker develops a complete and meaningfull job and has a clear vision of the influence that his work has on the final result. We aim at giving to him a decision space for autonomous organization of work both in the operations sequence and in the determining moments and frequency of actions directed to keep under control the productive process.
The long traditional assembly lines have disappeared and a new organization was made where workers working in parallel can develop

more meaningful jobs.

A second solution has been found in the development of multiscope robots. These systems are designed in such a way that by using different equipments they can cover a larger field of task than the previous traditional robots.

The first indication of the commitment made in this direction can be found in the agreement signed by Olivetti and the Metal Workers Federation in April 1974. To the comma concerning work organization we read: "In relation to jobs such as electronic assemblies, the assembling of groups and some workshop activities, where conditions for realizing significative professional enrichment by acting only on the technical side do not exist", job enrichment can also be reached diverting the assembling activity into activities of operating and controlling a specific mechanized or automated process.

Progress in the field of multiscope robots has made available robots such as the SIGMA (Sistema Integrato Generico di Manipolazione Automatica = Universal Integrated System for Automatic Handling) which has several advantages:

- the basic machine can be adapted to many operations;

- during the start up of new products the availability of the basic machine confines the design and the construction work to the specific fixtures only;

- the basic machine design costs can be spread over a greater number of units;

- construction costs are reduced because the machines can be produced in batches;

- maintenance problems are reduced;

- the basic machine can be paid over a period of time that goes behind the specific product life.

SIGMA has a high degree of sensitivity and it is easy to program because it can memorize operational instructions. These instructions can be introduced through specific commands which consist in accompanying the arm end along the track which we wish to be followed in the succeding operation cycles.

SIGMA has also an adjustment feedback which gives the possibility to modify and correct the operation cycle. In other words, SIGMA doesn't have a fix behavior and it has the capacity to insert the proper procedure whenever the actual situation is not identical to the expected one.

The system has sensors which can analyze different situations in a general manner eventually emitting a signal with univocal meaning and easy to understand.

We are not interested here to go deeply into the SIGMA's characteristics but rather to analize the consequences of the application of these machines on the organization and human roles.

A definite theory cannot yet be deduced from the few robots already installed for different uses, but we can begin to recognize definite trends which are sufficiently reliable.

HUMAN ASPECTS

As we already said, one of the aims of the new forms of work organisation is the professional enrichment of workers and the improvement of the work quality.

How this can be obtained through the use of multiscope robots it becomes evident examining how it has been applyed taking into account the fact that the most common applications are batch production oriented. For this type of operations it is necessary that fixtures be quickly interchangeable.

Within the fixtures we include:
- programs;
- handling devices and tools for specific operations;
- positioning and feeding equipment.

Programs

Robots requires an advanced programming work both in what concerns the operational cycle and the recovering procedures.

Such activities, which are now task of the tool engineer because the program is part of the project, could be lately assigned to the machine operator as his experience gradually grows.

Then it will be the operator, who after having acquired sufficient technical knowledge and experience and through proper training, will be able to contribute in selecting type and quality of actions required by emergency situations.

Machine Set Up

This includes:
- program loading;
- turn around of handling devices and tools;
- turn around of positioning and feeding equipment.

These operations too, after proper training, can be assigned to the robot operator.

Running

The machine operation includes:
- adjustments to unespected situations
- diagnostic and restart routines in case of machine stop downs;

- feeding of parts to be assembled and removal of groups assembled.

Operations such as adjustments, diagnostic and restart routine are activities that allow space for improving the professional level of the operator.

At this point we have to face a problem: feeding of parts and removal of groups, are activities of poor content. The solution to this problem is of high importance since the availability of sensors and recovering procedures could let the SIGMA function indipendently for relative long periods of time, resolving the bond of man's dependence on the machine. In order to minimize the feeding operations, we forsee two solutions that have in common the need to present pre-oriented parts to the manipulators, particularly if we refer to assembling operations.

The use of mechanical manipulators that would also orient the pieces could be possible but, at present, it is not advisable because they cost too much and they slow down the operation. On the other hand, manual pre-orientation is to be excluded because of its cost and its very low professional content.

Two solutions are possible:
- to provide specific preorientation equipment for every group or component (i.e. vibrating devices).
  In such a solution the increase of number of parts to be assembled causes an increment in the costs of specific equipments as well. Furthermore, orientation devices take up so much space that the number of parts to be assembled will be necessarily limited;
- load oriented components in special containers or racks to be attached to the robot.

The latter solution is probably the best because the equipment can be standardized and simplified; as a consequence its cost can be reduced. Beside, due to the low volume of the containers or racks a greater number of different parts can be assembled.

When electronic components have to be assembled on PC boards, for instance, undoubtedly this is the best solution because the components can be obtained from the suppliers already enclosed in a particular order inside special containers or racks.

The field of application can be extended by the use of standard, optical type, preorientation devices. The development of a standard orientation device at a low cost would definitely open the batch productions to automation. At present, the traditional solution, vibrating feeds at considerable cost and frequent re-adjustments, is applicable to the majority of cases. However, these problems can be somehow solved as explained hereafter through some examples of SIGMA applications.

Assembly of a capacitive modular key top (Fig. 1)

The two arms load on to the fixture, in the proper order and sequence:
- 5 percussion caps and 5 springs
- 5 "key body" and 5 pins

assembling 5 key tops at every cycle.
The machine time for 5 end groups is equal to 16". The machine can thus run for as much as 1200". The vibrating feed devices can be loaded during the machine time.

Assembly of micrologics on PC boards (Fig.2)

In this cycle the two arms must work at the same time, while one takes away the assembled board the other one replaces it with a new board.

At this point the two arms begin to work indipendently taking the components from the containers and inserting them in the board. The machine time for a board with 50 micrologics is equal to 120". (This includes the loading and unloading of the boards).
The SIGMA time range is about 3600". A buffer area allow the replacing of containers while the machine is operating.

Additional activities

Taking advantage of the extended time ranges the operator can operate more than one machine at the time (as in the case of micrologics assembly in PC boards) or he may perform additional activities such as testing and repairing the items produced. These last two activities are of definite help to the professional advancement of the worker because they allow him quite a degree of decision and self organization.

WORK ORGANIZATION

We have examined so far the aspects relative to the operator's assignments; we must now look at how the insertion of the SIGMA in the production process can be carried out.

A solution must be found in each case, after considering:
- annual volume/total volume of items of the various units that have to be produced;
- cost of the specific equipment necessary to adapt the machine to each type of unit;
- operational output/time unit (cycle time/number of units produced at every cycle);
- required assistance during the course of the operations due to feeding, etc..;
- possibility of assigning additional activities, such as testing and repairing, to

the operator.
We'll make a point going through two examples of the SIGMA application.

Assembly of a typewriter keyboard
Figure 3   One can see how the assignments are distributed on an assembly line of a traditional organization.
The arrows indicate those operations which are partially or totally automized.
In this case it is obvious that the introduction of SIGMA affects the whole cycle.

Figure 4,5,6 Show the SIGMA performing operations 1,2,3.

Figure 7   Shows the organizational layout after the SIGMA introduction.
The workers of this group are assigned an objective: accomplishment of an output, in terms of finished and tested units, according to the number of hours worked.
Such an objective can be accomplished through the performance of tasks that include both attendance to the machines and manual operations.

Assembly of the tape cartridge
Figure 8   Shows the organizational layout after the SIGMA introduction.
To the same worker can be assigned the set-up and the attendance of the SIGMA as well as of those machines necessary to complete the cycle (solder and testing) and the execution of all complementary activities.

PRODUCTION ORGANIZATION

As we have already said SIGMA is particularly applicable to discountinued operations typical of medium size productions.
Greater opportunity can be obtained through:
- a product design that takes in consideration functionality and the handling aspects of an automatic assembly even through a modularity concept. Modularity means not only a product broken down into separate testable units, but also and above all, the possibility of using the same group or the same component of a group to obtain the same function on different products;

- a very close contact between designer and engineering in order to make available tools and fixtures at the very beginning of the production period. In this way all the cost burden will be supported by the total number of unit produced.

- the development of organizational forms similar to the group tecnology seen from the assembly stand point. This type of organization would make possible the use of flexible automation systems with products having low production volumes that, taken individually, would not result economically sound.

At the same time, money could be saved on specific devices and start up machine time, through the use of modular equipment.

ECONOMIC CONSIDERATIONS

The decision to use or not to use robots is based on both technical and economic considerations. The main variables that influence the economic evaluation in comparison to the more traditional solutions (manual assembly) are:

C_R = robot cost;

C_S = specific tools cost;

C_L = cost of each second of worker's presence;

T_M = time in second to obtain a piece manually;

T_A = time in second of man's intervention on each piece produced by the robot, including the machine set-up time;

N = total number of pieces to be produced;

T_R = time in second to obtain a piece on the robot.

The point of balance can be determined as follow:

$$N\ T_M\ C_L \geq C_S + N\ T_A\ C_L + N\ K_R\ T_R \quad (1)$$

where $K_R = \dfrac{C_R}{3600\ H}$ where H is the total number of the machine working hours during its life time.

This last expression takes into consideration what has been already said, that the machine can be used for different jobs so that its useful life is not dependent on the product life.

In (1) T_M and T_R make possible a comparison between man's speed and the machine speed.
Such a comparison is not so easy because the definition of normal manual production rate is objectable. The so many attempts to define the concept of standard rate are all inaccurate because neither they don't avoid to take in consideration the operator's capacity nor they define such capacity in a uniform way.

As a matter of reference we will use G.B.Carson's definition (Production Handbook) where he says that standard rate is the time necessary to a normal operator, working in normal conditions, with normal capacity and speed, to complete a unit of work with a satisfactory quality.

Generally speaking, apart from any scientific definition of standard rate and basing our judgement on average working rate, we can say that according to our experience, in highly repetitive works, where few parts are repeatedly handled, SIGMA results at least three times as fast as man.

CONCLUSION

The main advantages coming from the use of multiscope robots are:
- the greater speed of the machine in comparison to the manual speed;
- the greater return due to higher productivity;
- the possibility of improving the work quality whenever professional enrichment for the workers, acting on technical aspect only, is impossible.

At this point one may think that there may be problems for future employment, but according to the writer there is nothing to worry about because:

- automatic flexible systems cannot be considered the alternative solution to the traditional work organization.

We have already explained how Olivetti, in order to counter the variances coming from the context in which it operates, has developed the U.M.I. organization. We are therefore very far from a completely automated plant. It is still necessary to encrease the robot artificial intelligence and to solve those problems of orientation and movement of parts among the various machines. Furthermore one should not forget that a batch production must be dynamically managed;
- the higher productivity of robots is a mean to obtain more competitive products on the international markets and therefore as a mean of aiming at higher production levels;
- and finally, with the extensive use of robots it is better to think in terms of job conversion rather then in terms of reduction of available jobs. The production of these systems will create new opportunity for employment in professionally significant activities such as designing, construction, programming and maintenance of machine and equipments, always keeping in mind that such activities require the integrated knowledge of hardware (mechanical and electronic) and software.

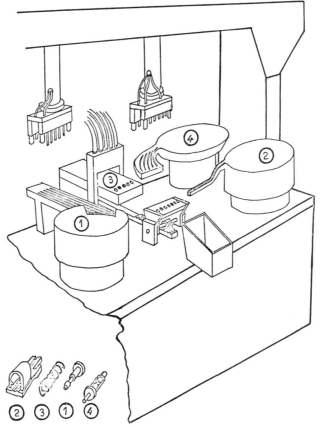

Fig. 1.

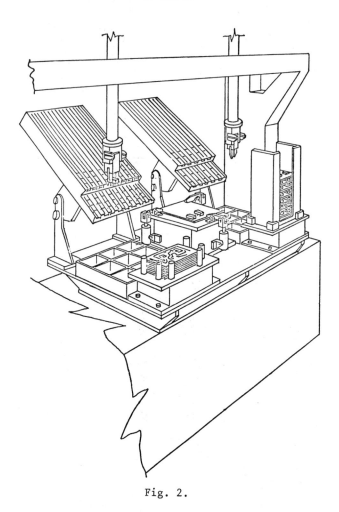

Fig. 2.

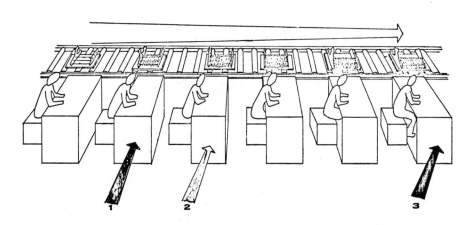

Fig. 3.

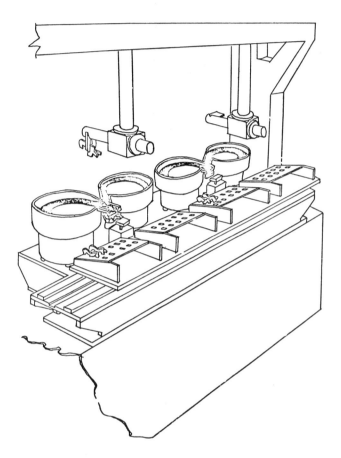

Fig. 4.

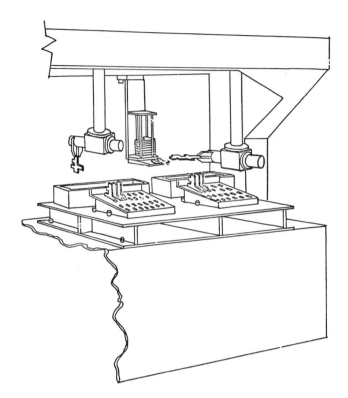

Fig. 5.

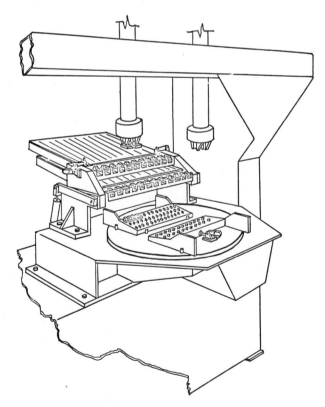

Fig. 6.

Work Organization with Multi-purpose Assembly Robots

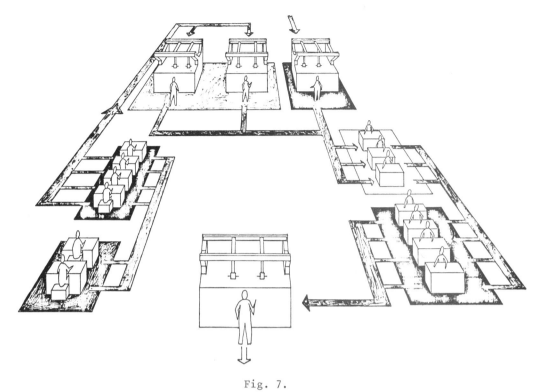

Fig. 7.

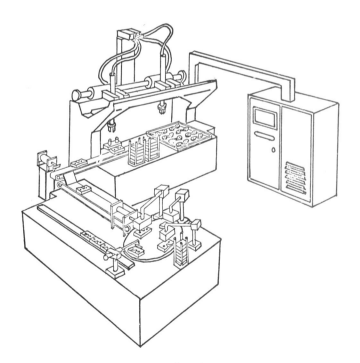

Fig. 8.

MAN-MACHINE INTERFACES IN THE CONY-16 INTEGRATED MANUFACTURING SYSTEM

Laszlo Nemes

Computer and Automation Institute, Hungarian Academy of Sciences, 1111 Budapest, Kende u. 13-17 Hungary

INTRODUCTION

At a time when production processes are subject to fundamental transformations, man - the main participant in production - sees his environment, his working and living conditions undergo rapid changes due to the accelerated progress of the means and structures of production. His reaction to those changes is a certain anxiety regarding his future prospects.

I think the present era is such a turning point. Computers, which were invented only three decades ago, are now fully fledged and have opened a new vista in industrial automation. The first rudimentary trials were reported some twenty years ago and since then computer control has achieved remarkable results in different domains of manufacture. New ideas have originated from the successes of the new techniques. In the most far-reaching concepts integrated data and material processing systems can cover the automation of the total manufacturing activity. The first experimental system have been successful, and the integrated computer control of production has gained ground in almost all branches of industry.

In both research and development a great effort has been made to improve these systems both in size and in complexity in order to perfect them in technical detail, to satisfy the requirements of flexibility and expandability.

The history of automation has set many precedents to emphasize the importance of the totally new aspects, when a high-level control system passes from the experimental stage into production. For wide range application the solution of scientific problems and the painstaking elaboration of technical details are indispensable but not sufficient.

Of the other considerations to be weighed, I should like to single out the social and human effects of highly developed production systems. This is an extremely complex question modulated by many crosseffects, the analysis of which is in itself a special scientific discipline.

Nevertheless it is those working in automation research and development who have to assume the prime responsibility for the new working conditions, for human part in automated production, to ensure that man is not a mere servant or a victim of technical progress. The duties to be fulfilled and the human interface to the production process must be planned in a civilised fashion.

However, the humanization of work is a rather inexact concept. Ever since primitive times man has found stimulus in creative activity. People of our times want to experience the same feeling. But working in more and more productive systems, they have been drawing apart from the creative process. It is the aim of the humanization of work to let people find their satisfaction even in highly automated factories. To be aware of being very useful to society, and to see their work being held in great esteem; this could be the meaning that people are looking for. Workplaces of the automated workshops should be conceived in such a way that the workmen should feel the importance of their work. This contribution to the success of production must be recognized. These considerations seem to be evident, but their realization is more than difficult. In those rare occasions where humanization of work was considered an important factor, several experiments were conducted but only very moderate results have been attained.

At the beginning of our research on the field of integrated manufacturing system we hoped we could adapt well developed ideas for humanization. We have come to the conclusion that this problem is still in its infancy and we must establish our own approach to solve it.

HUMAN TASKS IN HIGHLY AUTOMATED PRODUCTION

Before analysing human reactions and motivations in highly automated production systems, it is necessary to summarize the human tasks involved, which can be subdivided into three main categories.

The first group of activities concerns the jobs of controllers and operators. They are different in their complexity, but concerning the main activities they have many things in common. People in these positions are closest to the production process and therefore they are aware of their importance. However, the inadequate formulation of these tasks may often lead to an exeggarated feeling of responsibility so that they feeld inadequate and overburdened. A long training period is usually needed, since incorrectly formulated tasks require a very broad knowledge of the situation. Since an erroneous intervention may provoke fatal consequences and high material losses, after a certain period the work does not give any feeling of job satisfaction, but on contrary, engenders fear. People working in such circumstances might hold automation responsible for the stress they endure in their work. The second category of tasks can be better circumscribed: they are repair and maintenance less responsibility is needed and after a certain training period the sense of success is nearly certain if the task has been seriously considered during the design of the system. The more complex the system is, the more emphasis should be given to the automation of diagnostic and functional checks so that the repair and maintenance procedures should be prompted by the control system.

Machine operation and unskilled ancilliary work - though still extant - is endevoured to be minimised in the more recent systems, this being one of the most important arguments from the financial and sociological points of view for a highly automated production model.

I shall not endavour to analyse these three tasks in detail, instead I am going to deal with the human tasks most closely connected with the control system. Before discussing the various tasks of a controller-operator I would like to mention shortly the main characteristics of our CONY manufacturing system.

THE CONY-16 MANUFACTURING SYSTEM

It was during the realization of the CONY-16 control system, that we tried to analyse and subsequently humanize the different human tasks necessary to run the system. Our manufacturing system is small in size but it inclues the automation of the most essential fields of modern part production. It integrates material and data processing from the product design and production planning to the on-line control of the machine shop (Fig. 1).

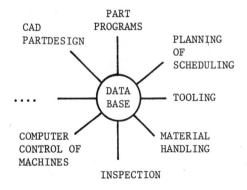

Fig. 1

This figure shows the integrated material and data processing system in the machine industry.

Since the desing programs consist of independent modules many variants for each workpiece can be computed in a short time. The planning and scheduling for economic production is also planned by autonomous modules. There are three main phases, arranged in a hierarchical organization; for production planning, scheduling and short term scheduling. The lower the level in the planning system the shorter time horizon is considered for the planning period and the more detailed definitions are given for the jobs.

To understand our basic considerations to be described in this paper the computer control of the machine shop needs a somewhat more detailed explanation.

The complexity of the computer controlled machine shop is characterized by the relation of the production-cell with the outed-world (Fig. 2). The raw materials to be loaded into the production system are delivered in batches at certain intervals in accordance with the available supply, therefore an input buffer is needed. The machining of the parts stored in the buffer proceeds according to the day's production plan.

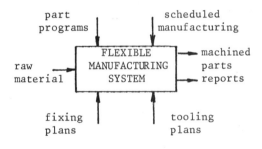

Fig. 2

The main interfaces of a flexible manufacturing system.

Should any disturbance stop the flow of production, the information collected must be used to detect, and circumvent the fault and to draw up a new production program. The tool and jig plans and the geometric and technological data controlling the programs of each operation complete the initial information, which all define what we call briefly "HOW" and "WHAT" to produce.

The tasks of the production system shown in Fig. 3 are grouped according to the various kinds of operations each realized by computer. Taking metal cutting as an example, the computer controls the machine tools through a special peripheral (MTC). The worker operating the machine tool has from time to time to enter into a bidirectional communication with the computer. To solved - we had the choise between the following possibilities: to install commercial I/O equipment like displays or teletypes next to the machine tools, or develop MTC units including efficient human interfaces. It is clear,that we have to face similar decisions with the other items of Fig. 3, such as part changing, tool changing, storing workpieces and assembling tools.

- Metal cutting
- Part changing
- Tool changing
- Cleaning
- Inspecting
- Fixing the parts onto technological pallets
- Tool setting
- Part buffering
- Cleaning the parts of chips

Fig. 3

Technological tasks in flexible production system

One might argue why there should be a need for high-level human intervention at the workplaces of computer-controlled production systems the reason is that the preparatory phases are open loops and so technological errors and any mistakes concerning the production equipment and the hidden defects of the raw material can be perceived only during machining. Now it depends on the ability of the operator wether those abnormalities can be detected in time and their influence minimized. In the majority of cases several kinds of failure can be diagnosed by the computer, but certainly there are and there always will be situations where it becomes the duty of the worker to stop operation and supply the central control system with sufficient information concerning the abnormalities present.

The supervisor is charged with the surveillance of the entire machine group and is therefore in great measure responsible for it. In order to maintain a continuous operation of the very costly production system; he has to oversee the vital parts of the system, that is the circulation of the data and material processing, its rythm and its errorless performance.

He has to be informed about the cutting processes and about any fault during workpiece change, buffering and tool change. He has to know if the ancillary human tasks have been carried out in accordance with the production schedule. The machine group supervisor has to perceive any abnormality in production within a very short period, he must be able to decide on the effects and size of the problem and he must intervene quickly in order to restore production. This work includes also various other kinds of duties; communication with the operators, with the control software system, the making of decisions and giving orders. We are certain, that a supervisor has to be a technological expert and not a computer buff.

MODEL OF THE CONTROLLING-OPERATING JOB

In order to ensure the creation of optimal circumstances for the most critical and responsible jobs in highly automated production systems, it seems desirable to construct a model of the human activity involved. Three qualitatively different sections can be distinguished here, for the composition and realisation of a humanised task. We decided to analyse the three phases of this model separately and to define the tasks and working methods accordingly. Let us now consider the phases in simewhat more detail and separately.

The first phase of these tasks is information gathering.
It is very easy to say that the information available should be essential and sufficient. However, this means that a careful and constant analysis is needed. This is necessary firstly because at all points of the process information is generated which might be needed by the operator or the controller. But if all this vast quantity of information was presented to the operator or controller by means of displays, he would be faced by an almost impossible task. He would be expected to master an immense and therefore incomprehensible quantity of information and this would put him into a nearly constant state of uncertainity. He would realize that he is no more able to survey the entire process, thus the control of the process will slip out of his hands and he will feel defenceless against events. For this reason we have designed with special care the information about the course of the process. This began with the choice of the information device, that is incorporated in the hardware. We decided to abstain from displaying codes, having lamps blinking and have presented the information always in text form, in all cases with the numerical values belonging to it.

We have chosen alphanumerical displays even on machine operators tablets, and the display of 20-40 characters has seemed to be sufficient.

Messages consisting of several parts are represented sequentially in the order of their importance and the receiving and comprehension of the message must be acknowledged by the operator. In this way he will build up a feeling of security, that nothing was communicated, no information was generated without him being informed about it and taking note of it.

At the control workstations we have chosen a more complex information tool, the full CRT alphanumeric display. We have very seriously considered the choice of information to be presented on the screen of the display. It

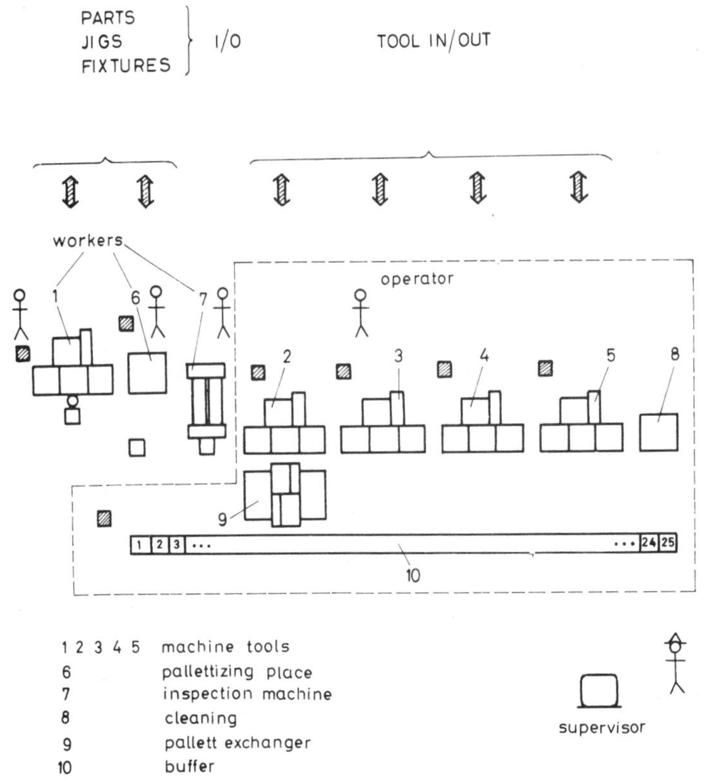

PARTS
JIGS
FIXTURES
I/O
TOOL IN/OUT
workers
operator
1
6
7
2
3
4
5
8
9
10
1 2 3 ...
... 24 25
1 2 3 4 5 machine tools
6 pallettizing place
7 inspection machine
8 cleaning
9 pallett exchanger
10 buffer
supervisor

Fig. 4.
Flexible manufactoring system controlled by cony-16

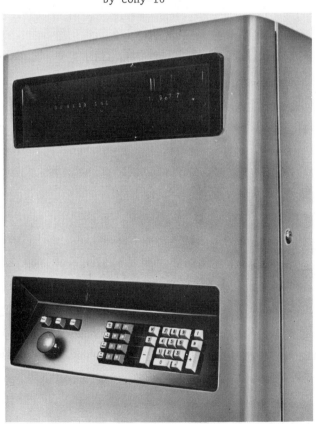

Fig. 5.
The controller unit

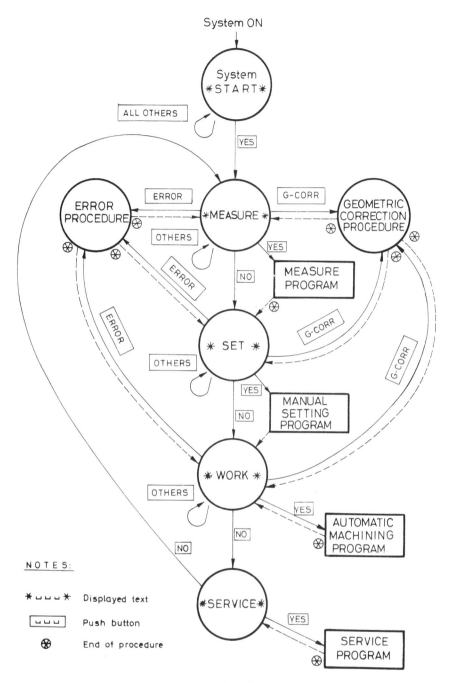

Fig. 6.
Basic dialog context structure

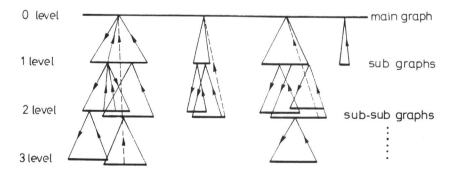

Fig. 7.
The organization of the dialog graphs

has been a basic principle also here, that meassages of information for monitoring should only appear in the sequential form. At the formal request of the controller, however, partial results of the process can be also requested in a tabular or summarized format. These more complex tables of information, however, are to be considered as supplementary, and never as the automatic messages of the process. Thus the degree of detail of overwhelming quantity of data and the controller will request further information only in such cases and upon such problems, where it becomes necessary for the performance of his task.

Decision is the second phase of our chosen model,
and here a judgement is formed which corresponds with the actual needs of the given situation and is based upon experience or knowledge. We consider, that the mental process, connected with the decision, is the most essential responsibility for an operator-controller. Thus the task of decision-making must be entrusted to the human operator and as far as this decision is supported by broadly based and programmed information, the operator or controller will increasingly feel his work to be creative and diminished in quantity. Of course, the appropriate empiric and learned knowledge is necessary for the ability of making decisions. This makes a long period of training and learning indispensable. However, the process of decision can be assisted without taking from the operator-controller the responsibility, which is the human creative part of the work.

Essentially we should formulate, that the hierarchy of the decisions should be offered to the operator in the proper time and situation, perhaps in the order of their probability, and if he has aquired enough information about the process, then he will in fact choose amongst all possible variations the optimal one for the given situation and he will not decide according to the first idea which comes into his mind. Information and decision are in many cases not necessarily a sequential process. In the case of a thoughtfully prepared and really humanised workplace the appearance of the decision-variants will require some further informations and therefore the process of information and decision may indicate an interactive relation. The process must be terminated by a definite decision and if the computerized background had adequately supported the information and preparation of the decision, then - according to our experience - the feeling of fear of responsibility has disappeared and the decision is based on a feeling of being the master of the process.

Intervention is the third phase of our model.
The requirements of intervention can be formulated by simple words of general validity, e.g. that intervention should be rapid, should not need long training and should, if possible, exclude errors. Interventions and series of interventions have, of course, their rules.

In order to start functions and functional tasks the intervening person has to learn some semantic rules to be able to attain the effect he had in view. We have formed the opinion, that the intervening operator should, if possible, not have to memorise the intervention, the possibilities and the way of doing it. Thus we have reduced intervention to the two sub-processes of information and decision, where in the information phase the operator is instructed by his panel about the possible ways of intervention and decides in the subsequent sub-period (of decision-making) between the variations of the intervention. The ways of intervention can be chosen relatively rapidly and easily by this method and a high-level operating technique can be attained in a very short time.

According to our experience, this model of operating-controlling work, consisting in fact of information gathering, decision making and the subsequent intervention, affords very cultured working conditions for the human being, and this is even more so, if the computer controlling and directing the process automates the information-decision and intervention to the best possible degree. The previous methods of learning syntactic and semantic rules for systems operation are not required, it almost appears that these are taught to the operator by the system.

The example of realisation.
In the case of the manufacturing system CONY-16 the basic principles were elaborated by uniform consent, but there arose different opinions during the choice of the technique. We have initially supposed, that a computer controlled dialogue-system can be most favourably employed for the service of all three phases of the operator-controller model. A computer initialized dialogue-system presents an adequate guarantee that the operator cannot forget anything in any phase of the model, avoiding at the same time, that detail problems should require his attention thus omitting vital parts. The counter argument was brought up, that a principally pre-programmed dialogue-system does not humanize the workplace but rather mechanizes it. Even the argument was heared, that thus human capability of control and initiative will cease to exist. The operator - instead of arriving in an intuitive way and within one step to the final result in any phase of the model - is obliged to search the solution in several sequences.

Well, initial experiences have clearly demonstrated the correctness of the development concept. People like to work on the dialog-oriented operator-controller's workposts. They are handling them after relatively short training periods and they do feel, that the informative and intervention organs in the system are indeed at their service. The experiences also showed that operators and controllers are convinced that the computer not only controls the process but helps them in their work and by performing the routine tasks of their job, their energy and concentration capacity is liberated for the decision processes which they regard as creative work.

The CONY-16 manufacturing system and its man-machine interfaces.

The machine shop area is shown in fig. 4. The four milling machines (2, 3, 4, 5), cleaning station (8), inspection machine (7) are served with automatic part changer, rolling back and forth along the manufacturing line. On the opposite side, of the changer there is a buffer with storage capacity for 25 workpieces.

The base surface of the raw material is machined first (2). Next, the part is fixed on a technological pallett. The pallettizing place must provide enough assembled palletts during its eight our working period, for the remaining part of the manufacturing system operating in two shifts. From this point the changer carries the pallett to the machine tools or other stations according to the process chart automatically or puts them into the buffer if the required technological stations are fusy.

After machining the parts are cleaned (8), than measured on the inspeciton machine (7).

The base surface cutting machines, the pallettizing place and the inspection machine have their own operators who work in almost the traditional way like working on separate machines, but all of them need communication facilities to receive and to send messages to the central computer.

The four machines, part changers and buffer are served by one person whose task is to load the automatic tool changer from time to time, according to the scheduled program. His job is, moreover, to make geometrical corrections for the part program, according to tool wear and inaccurate tool setting. In the case of any irregularaty he must intervene, inform the computer and/or supervisor, and with their active assistance he locates the error and helps to bring the system back into normal automatic operation.

There are many types of errors: mechanical ones (tool wear or break, material defect), part program mistakes, or electrical failures. The operator-computer interface, both in hardware construction and in software assistance, must offer an efficient help in these cases too. To facilitate, these activities a speacial operator interface system has been developed.

The operator interface

The equipment, illustrated in Fig. 4 as shaded squares and is shown Fig. 5. In the upper window there is an alpha-numerical special readout panel to display the messages from computer and the supervisor. The input device is a simplified keyboard divided into three main groups: numerical keyboard, functional keyboard and special keys for answering the dialog system. A more important part of the operator-system interface is the software for the degree of humanization takes shape here.

The structure of the software is organized in the form of dialog conversation. The operator receives prompts from the system and he must make his own decision of accepting or refusing them. In the latter case he would be offered the next choise for the next decision.

In case he accepts the proposition he would be asked for more detailed information possibly for numerical values. The global graph structure is shown in Fig. 6.

After the system is switched a housekeeping procedure is done automatically whose end will be notified with the xSTARTx message on the display. In the START position any button - except the YES - can be pressed without having any transition, and this state will remain until the YES button will be pushed, when the xMEASUREx message will appear on the display, offering this mode to the operator. He has a choise of refusing it (pressing the NO button), accepting it (pressing the YES button), or of making a digression from it by entering either into the GEOMETRIC CORRECTION procedure (by pressing the GKORR button) or into the ERROR PROCEDURE (by pressing ERROR button). In the two latter cases the control of the program will be returned to the xMEASUREx state automatically at the end of either procedure.

In the case of accepting the xMEASUREx mode an other dialog graph of the MEASURE subprogram will be activated offering all the possible of the MEASURE PROGRAM the xSETx mode will be offered automatically. On the hand of the measure state the next mode xSETx will be put forward. From this point onwards the game is the same, as was shown in the previous example. The rest part of the figure does not need further explanation.

In the foregoing we introduced the hardware and software means which establish a high level operator-computer interface next to the machine tools. Now it is possible for the operator to solve his problems in the manner in which factory people like to express theirselves.

The main question was to define the domain of activities which will be the subject of the software, utilizing all the previously described aids. More precisely the question at this stage is how detailed graphs are necessary at certain levels, and how many levels are needed. (Fig. 7) We have figured out, that an effective dialog system for manufacturing processes should be designed so that
- each separated graph should have max. 5-10 nodes,
- the graphs should be organized in hierarchical structure
- the more frequent operation should be placed on the higher level,

- the more frequent operation on the same level should be near to the entry point of the sub graph
- the most frequent operations must be organized into small graphs and can be nested into max. two levels
- sometimes it must be ensured, that a return edge of a graph can be pointed to the higher level but next.

CONCLUSION

The first experimental system of CONY-16 was built on a small scale a few years ago just to test the basic ideas in factory environments.

The people who were trained to be operators mastered running the system within a day. In comparison the training for handling conventional NC machines or computer peripherals used as interfaces in the workshops needed much longer training periods.

After finishing the factory test we dismounted the experimental system. The workers protested, demanding a new computer controlled manufacturing system which makes their life easier and more comfortable.

AUTOMATION AND WORK ORGANIZATION
— an Interaction for Humanization of Work

Jan Forslin

The Swedish Council for Personnel Administration

Traditionally, the question has now since many years been put roughly this way, "how will work be affected by the (inevitable) process of automation?" The answer to such a deterministic question has only recently become "it depends", or rather the consequences will be those you want to have. That is, contrary to what was formerly thought, for the same technological change there is a choice to be made in terms of what working conditions you want to create. It is mainly an organizational choice, e.g. given a certain technology, you can organize work in many different ways, choose different solutions maximizing different values. Such values can broadly be seen as economic, technical, human, or social. By using two case studies I would here like to illustrate the great importance of work organization in connection with automation for the impact on humanization of work itself.

In the first case one was not aware that organization is a relative variable, subject to deliberate alterations depending on your assumptions about human nature. For this reason, the opportunities both of getting a more efficient work process and of making work more human were missed in almost a whole branch of industry. In the other case the options of different types of organization had been identified, and this opened the possibility, in connection with a new technological step, to abandon a destructive tradition.

The first study deals with the rapid and drastic technological change in printing industry that transforms composing rooms from hot, dirty machine shops into cool, elegant office landscapes. Gone are the lead types, the noisy setting-machines, the sawing and cutting in metal. Instead there are smoothly ticking computers, high velocity photo-setting machines, viewing screens and paper strips. Gone are the honourable occupational traditions of a workers' élite, in many cases substituted by temporary unskilled labour.

From the workers' point of view this broad and far-reaching technological change, that does not leave a thing standing, was correctly perceived as a rationalization, threatening the strong position of the occupation and even likely to cause unemployment for maybe the first time in their history. The trade union has consequently concentrated its efforts on securing the jobs for their members. And as the physical working conditions were so spectacularly improved by the change, not much notice was paid to psychological aspects of work in the new technology.

The study was undertaken in, according to Swedish standards, medium-sized newspaper plants with approximately 30 persons working on the technical side. In these cases, technical management consists of the foreman. He is normally an autodidact specialist with long practical experience of the traditional technology on whose shoulders the burden of making a successful switch in technology rested. The change in equipment had, literally speaking, to be made overnight, thus causing a lot of stress on the foreman. He had to rely on his own judgement in the choice of equipment and in the technical solutions. Much money was at stake and a mistake in the change-over would cause the embarrassment of the paper not coming out that day. Actually, several of the foremen suffered from heart attacks during the preparation period. So in this situation of high insecurity, the foreman wanted to draw on already tried and approved modes of production. This meant that as much as possible was imitated from the old technology. There was a very specific translation of jobs in the old organization into the jobs in the new, trying to preserve as many characteristics as possible in spite of the technologies having almost nothing in common. There was thus, neither from management nor from trade unions, any strong incitements to search for new solutions.

This had the unfortunate effect that the bad characteristics of traditional setting work were preserved and the opportunities the new technique brought in its train were not recognized. Contrary to what is often thought among social scientists, the skilled and almost intellectual work of a compositor is not very interesting or rewarding. The only ones who did not seem to be sentimental about an old and self-confident craft disappearing were the compositors themselves. They wanted a less tiring and stressful job and they welcomed the change.

As can be seen from figure 1, in most important respects the working conditions in the traditional technology were unsatisfying, e.g. did not correspond to the lowest demands of the workers.

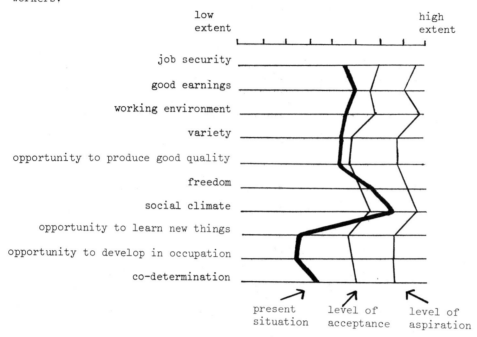

Fig. 1. Evaluation of job of compositors in traditional technology

Freedom in work and social climate supersede what is here called "limit of acceptance". Of particular interest here are the big deficiencies in connection with the true human aspects of work: opportunities to learn new things, to develop in one's profession, and co-determination - aspects corresponding to qualities differentiating men from machines. The occupation of the compositor can be said to be qualified but with not possibilities for growth and personal development.

In these respects there were expectations regarding improvement in the future as the companies were assumed to change technology sooner or later. As compared with the present opportunities the expectations are considerably higher for the physical working conditions and possibilities for personal growth.

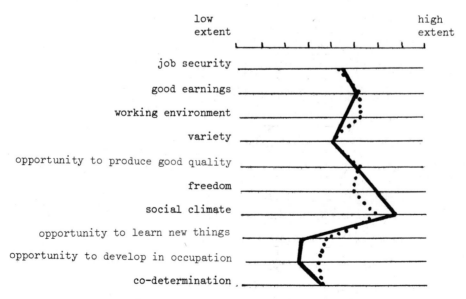

Fig. 2. Job expectations of compositors in traditional setting compared to their present job situation

Before going any further, let us take a look at some characteristics of newspaper production and the implications of the change in technology. Very roughly, the chain of production contains the following steps corresponding to the metamorphosis of the product from the idealistic formation of the content to its physical mass-produced counterpart.

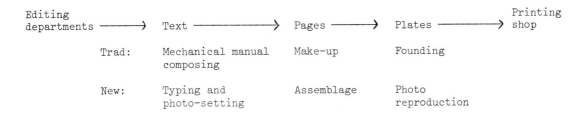

Some of the changes in the production in connection with the new technology are summarized below.

	Traditional technology	New technology
Technology	mechanic	→ electronic
	slowly developing	→ rapidly developing
	partly mechanized	→ partly automated
	flexible	→ rigid
Organization	embryonic line production	→ integrated flow line production
Material	lead	→ paper and film
Product	early visibility	→ late visibility
Technical competense	compositor and foreman	→ external specialist and foreman
Craft knowledge	compositors' training and experience	→ computer memory
Training	long apprenticeship	→ little training

How did the workers then perceive the changes; did these meet their expectations? When comparing the ratings of the workers in traditional type-setting with the ratings of those working in the new technology we can note improvements in many respects of the work situation.

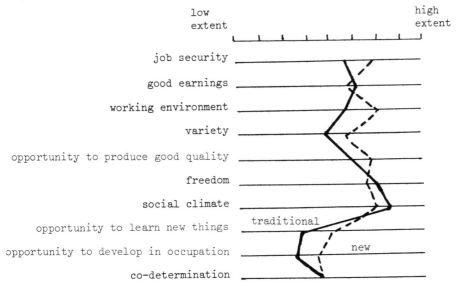

Fig. 3. Perception of job of compositors in traditional and new technology

The security is higher, as the problem of redundant personnel has been settled. Also the possibility for the newspaper to survive is seen to be increased with the modern technology. The already mentioned improvement of the physical working conditions is strongly reflected in this picture. In a number of psychological apsects of the work, the situation is also better: in general aspects connected with personal development - a previously critical area. There is one important exception to the favourable picture, namely the "social climate" has become less good, as the new technique has broken up old social relations among the workers. What has not been affected by the change in technology, are aspects regulated by personnel policy and collective agreements: earnings and co-determination.

Are these improvements satisfactory? If we add the aspirations of the workers to the picture we have a better ground for evaluating the new situation.

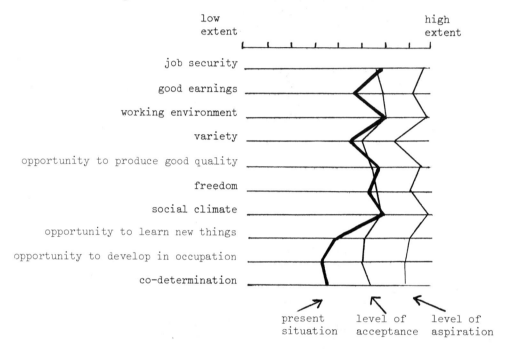

Fig. 4. Evaluation of job of compositors working in new technology

The security has now been raised to an acceptable level, as have the physical working conditions. The quality of one's work has also been raised to a desirable level. On the negative side is the circumstance that the previously satisfying social relations are hardly acceptable now. Finally, it should be pointed out that the opportunities for personal growth are still, in spite of improvements, far from satisfying.

Neither does one expect any drastic improvements in the future, as can be seen from the next graph. The expectations are sooner pessimistic than optimistic, but in general there are few expected changes and the situation is seen as stabilized in its new form.

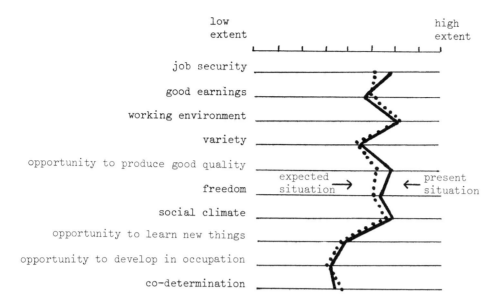

Fig. 5. Job expectations of compositors in new technology compared with perception of present job

So far we have seen that the work situation improved in connection with the new technology and none of the typographers wanted to have the old technology back. What more can you ask for? What are you as a social scientist fussing about? As I said initially, the risk in connection with technical changes is that you underutilize the possitiblities for improvements rather than create a less favourable situation than before. So let me point out a few issues in this case where the opportunities were not fully utilized.

There is a direct connection between, on one hand, the type-setters' needs for development in their profession, to grow in competence and to exchange their old craftsmanship for a new, and, on the other hand, the need of the production for a flexible labour force competent in handling all the different jobs, technical experts for quick problem-solving in case of breakdowns, and, the new technology being a fast-changing one, also workers open to future changes. This possibility of linking these two sets of needs has been lost by not choosing a group technology, by not delegating technical responsibilities to the type-setters and, for example, involving them already in the planning and designing of the new system. A group technology would also have created a better social climate where one co-operates, solves problems together and where one can develop close friendship relations.

Some of these mistakes can be compensated for in the future by reorganization of the unit. A broader challenge lies, however, in the possibility of tearing down the inter-departmental organizational borders altogether. Historically, it has been necessary to isolate the noisy and dirty composing shop from the rest of the organization. This motive does not exist today. The composing department is the critical link between the editorial parts and the printing shop. Here mistakes in earlier steps are corrected and adjustments are made to make possible the final step of mass production. There is constant miscommunication and bad feelings between the editorial staff and the type-setters causing inefficiency and frustration. These problems could possibly have been avoided if an organization had been chosen that integrated the type-setting with the editing. This would also have created more interesting work for the type-setters.

The second case is collected from automotive industry. Here the setting was quite different. The department studied was just one unit within a huge plant for engine production. The enterprise was very resourceful with internal technical staff departments and social scientists. The central corporate personnel policy was very explicit concerning humanization of work which exerted pressure on local management. Locally, there was also felt the pressure from below in terms of difficulties in recruiting, absenteeism and also a conscious trade union policy backed by experiences in the same branch of industry. An intensive management development programme aiming at middle management, corporate and local seminars for production engineers created a climate favourable to experimentation.

When a new major part of the plant was designed, a radical break with traditional work organization was made. The organization was developed simultaneously with the equipment and in close cooperation with the workers, the latter being a core group picked for the initial manning of the new department and later used as instructors for the newcomers. Previous experiences had shown that changes in organization implemented from outside, developed by specialists, had not been successful in other parts of the plant.

The unit is quite small, employing 22 persons in each of two shifts. As compared to the until then prevalent technology, the new department represented the next step in automation with the whole transfer line interconnected, the engine block automatically moving between transfer units where also loading and unloading of the unit took place without human assistance. The new department, in comparison with traditional units with similar production, shows many differences regarding work organization. The new technology required and made possible a new work organization. It would have been equally possible to apply the traditional organizational principles with "one man serving one machine", but instead the group principle was agreed upon as a basis for the organization.

Within the groups the members have shared responsibility for keeping the machinery running and they have been given much freedom to organize their own work. Quality control is also - to a considerable degree - left to the worker's judgement and not - as a general rule - regulated by detailed control instructions.

Further the sharp bounds between different work roles, which characterize the traditional organization, have been abandoned. To give an example: inspectors are participating in adjustment work, or the adjuster is doing inspection work, depending on actual work load.

One goal is that every employee shall be given the opportunity to learn all tasks within the whole department. Besides the spontaneous learning taking place within the groups, there is a systematic planning for further training of every employee. In addition to on-the-job-training, there have been formal courses on machining techniques, tools and quality control.

The workers are paid on an incentive basis. The wage is based upon final output from the whole department. This means that the individual worker for his income is dependent on every other's effort in both shifts. This differs from the ordinary procedure in the company, according to which the workers are paid individually. In practice, income variations between wage periods are low. Full pay is received when output corresponds to agreed achievement standards and delivery programmes. It is not possible for the workers to increase the pay by producing more blocks than demanded.

ORGANIZATIONAL CHARACTERISTICS

	Traditional units	New unit
Work instruction	Close and detailed	Only for guidance
Being tied to a workplace	Yes	No
Dependence on machine pace	Yes	No
Operator's tasks	Simple operations on machines	Process monitoring, maintenance, control of dimensions, exchange of tools and setting up machines, trouble shooting.
Work role	Strict work roles	Flexible work roles
Work cycles	Short, 1-5 minutes	Non-cyclical
Social organization	One man - one machine	Group organization
Wage system	Individual incentive system	Group incentive system
Specialization	Operators highly specilized on simple tasks. Variations handled by specialist functions.	Broad competence in operators. Specialist work successively delegated to the operators through continuous learning
Communication	Strongly restricted possibilities to move geographically and to make contacts, also loud noise.	No geographical restrictions. Good possibilities for spontaneous contacts.

Automation and Work Organization

	Traditional units	New unit
Training	Very short period of training. No basic training for the job and no increase in competence.	Basic courses in production technique. Continous learning on the job.
Job rotation	No	Yes
Lay out	Line operation flow	Line operation flow

When evaluating the effects of these innovations, I will here restrict myself to the workers' reactions, as the main topic is humanization of work.

The basis of the record of the workers' reactions is a comparison between, on one hand, four units with differing technology but with identical and traditional principles for work organization, and, on the other hand, the new unit with advanced technology and radically deviating work organization. The rational for this is the wish to illuminate the effects of work organization supported by the fact that there are very small differences within the first group of units and, where differences exist, they are not systematic.

The effect of the changed work organization is clearly seen in the workers' ratings of the work content. The ratings by the workers in the new block department are more positive for practically all aspects. The smallest differences concern responsibility and independence. The degree of responsibility is rated high in all departments but still somewhat higher in the new block department. This probably reflects that all workers regard their work as demanding responsibility.

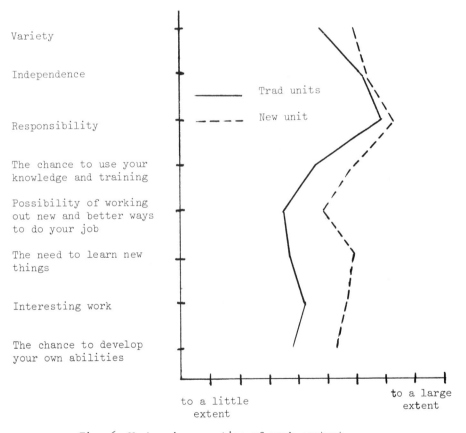

Fig. 6. Workers' perception of work content

The workers in all departments also consider themselves to be independent in their work in spite of considerably different degree of freedom in their tasks. For example, the workers in the new department are much less dependent on specified work pace, specific set of routines. The workers are also less tied to the work place and the work is less structured. A possible explanation of the similarities in the ratings can be found in the workers' interpretation of independence, specifically meaning independence of supervision, all work requiring very little, if any, close supervision.

The differences are particularly great in aspects related to personal growth and learning at work. The possibilities for self-decision on and development of work methods and for use of one's abilities and knowledge are seen as much better. Work is also seen as less monotonous and more interesting. The general picture is thus highly congruent with the values behind the change in work organization.

One important feature of group technology is of course the need to cooperate and to solve problems in groups. This has also been regarded as a favourable aspect of work, relations with co-workers having strong motivational effects. This group organization is reflected in the workers' ratings: the possibilities to talk with each other during work, as well as to help each other are rated better.

These improvements also depend on the lowered noise level and the lessened geographical confinement. The more personal relations are, however, not affected. The personal interest, which one takes in each other, is the same in all departments.

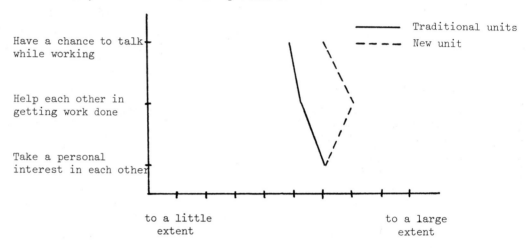

Fig. 7. Workers' perception of social relations

Great efforts were made in construction of the plant to minimize negative physical working conditions. According to the ratings, such conditions have been much improved. Compared to other departments, this unit is less noisy, has better air conditioning, the premises are light and spacious. The work is also seen as less physically demanding. The psychological strain, on the other hand, is seen as about the same. The risks of accidents are also seen as slightly higher, probably depending on new and complex equipment and a fairly large proportion inexperienced workers. These results are partly effects of modernization but also strong intentions to improve the physical conditions.

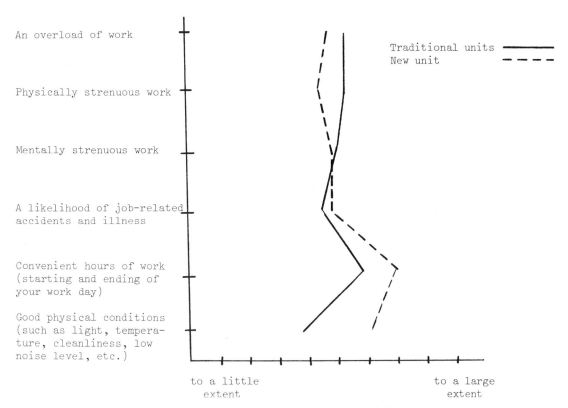

Fig. 8. Workers' perception of working conditions

The workers also rated the importance of the same aspects of work. The importance can be seen as a function of the aspirations of the workers, thus partly reflecting the strength of their needs. The ratings of the importance of the different aspects of work content should reflect the workers' aspirations in these respects - their need for self-actualization.

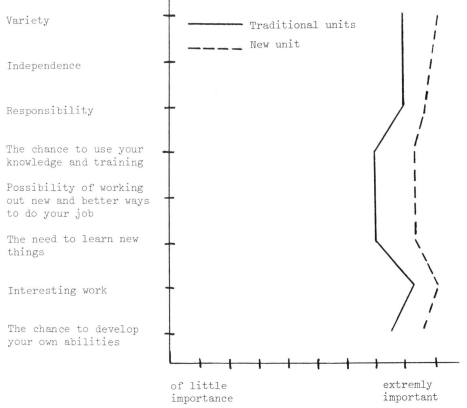

Fig. 9. Importance attributed by the workers to work content

The workers of the new block department attribute more importance to all aspects of psychological work content. Indeed, these are important to all workers, but even more so to those who have experienced a different work situation. This would then indicate that the basic assumptions about human nature and needs behind these changes in work organization are supported by data. The need for growth is increasing when you get a chance to fulfil it. The aspirations of the workers are rising as a consequence of better opportunities.

The rise in aspirations also shows in the ratings of importance of employment conditions. The ratings are about the same in all departments for pay and job security - that is of equal importance for all workers. But the growth-related aspects of employment are consistently higher for workers in the new department.

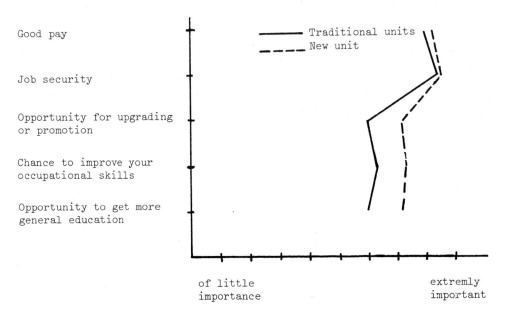

Fig. 10. Importance attributed by workers to employment conditions

As we have seen, great importance has been attributed to the workers' opportunities to grow and develop in his work. The conditions for this are provided by the work organization and learning in the daily work, as well as more formalized education in the form of shorter courses.

It is important to underline, though, that there are no formal educational requirements in addition to what is provided by the school system. Also the smallest training required to be able to participate in work in the new unit, is very short and does not deviate from that of other units - normally up to one month.

It is now reasonable to ask whether these generally more favourable ratings by the workers from the new department depend on a positive bias like a Hawthorne effect. Or is the social climate so good that the answers will be positive irrespective of what is being asked? It has already been shown, however, that there are deviations from such general patterns, and this is even more evident in the following graphs.

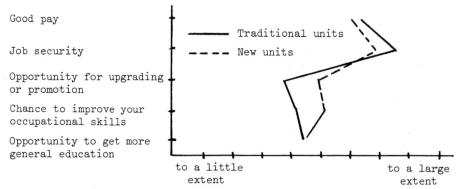

Fig. 11. Workers perception of employment conditions

Here it can be seen that pay and job security are rated less favourable in the new department, probably due to the fact that the pay system had not yet been fully agreed upon and that at the time of the investigation some transfers had to be made. Again, the conditions connected with personal development were rated higher, with one important exception: The opportunities for general education have not been affected by the change in work organization. This shows clearly in ratings which are not different between departments.

The next figure demonstrates a very clear case of consensus on certain values between departments. In spite of very different working conditions, the ratings from all departments coincide when the importance of these aspects are being rated.

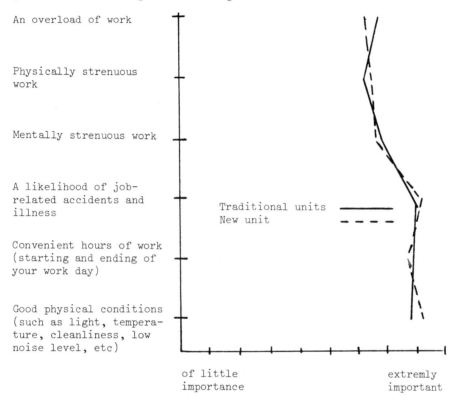

Fig. 12. Importance attributed by workers to working conditions

This is further evidence of the validity of ratings but it also reflects an important motivational difference between the psychological content of work and the physical working conditions. Herzberg would differentiate them as intrinsic and extrinsic factors. If the psychological content is improved, then the aspirations for personal growth increase, but improvement of physical working conditions does not have a similar effect. Removing of physical nuisance in work is appreciated, but does not create a need for further improvements.

FREEING THE OPERATOR FROM THE MACHINE—
AS EXEMPLIFIED ON ASSEMBLY LINES IN THE
AUTOMOTIVE INDUSTRY

Rolf Knoll

c/o Daimler-Benz AG, D 7000 Stuttgart 60

1. INTRODUCTION

The producers of mass consumer goods - also including automobiles - today have to comply with two requirements which on first glance are contradictory or at least difficult to combine.

On the one hand they should apply the <u>most rational production method</u> to allow the supply of the market with products priced as attractively as possible. On the other hand they should offer <u>safe and secure</u> jobs to their employees, which at the same time should be interesting and require no heavy physical exertion. Today this falls under the general heading of <u>"humanisation" of working conditions</u>.

Both demands can be justified when placed in reasonable relation to each other. Therefore it is the job of the management and its specialists to produce solutions meeting both demands and put them into effect.

You all are familiar with the development of highly automated manufacturing facilities as we have been using in the manufacture of components for passenger-cars such as engines, transmissions and axles since the mid-fifties. Here we have succeeded to a great extent in freeing the machine shop worker from repetitious, hard, manual operations in favor of using his capability to perform monitoring and controlling functions.

2. MECHANIZATION OF ASSEMBLY LINES

In the assembly plants automation progressed at a much slower rate. The sequence of motions executed by a worker in the assembly process is much more difficult and expensive reproduce by machine. Initially, therefore, usually manually operated multi-spindle power wrenches, insertion tools, or handling devices were utilized. The more complex and the larger and heavier the equipment the more difficult was it to manipulate. The next logical step therefore was to integrate this equipment into the assembly lines as automatic assembly stations.

As a consequence of this concept, however, the remaining manual operations were frequently forced into the inflexible sequence of the automatic assembly stations. This meant in some cases that the assembly worker had to adapt his rhythm to the machine, far more than on the strictly manual line earlier. Moreover, he lost the personal contact to the workers next to him. Thus the problem arose of disengaging the worker from the assembly machinery.

For the sake of completeness, it should be added that, of course, a number of other parameters had to be considered and improvements introduced simultaneously where possible, to name only a few of these:

- ergonomic design of working places
- saving of labour, job enlargement, job enrichment
- opportunity for job rotation
- increasing operating safety

I will, however, not go into further detail on these points.

I shall rather concentrate on showing, by practical examples, some approaches to the solution of the problem of freeing the worker from the machine cycle as they were applied in our passenger car axle assembly. We believe that these solutions took both, the humanitarian and the economic requirements into account.

3. EXAMPLES FROM PASSENGER CAR AXLE MANUFACTURE

3.1 Assembly of semi-trailing control arm for rear axle

In this procedure the semi-trailing control arm, rear axle shaft flange, tapered roller bearing, rubber bearings, brake carrier and various small parts are assembled and the play of the wheel bearing is adjusted and checked (picture 1). As part of the chassis section of the rear axle this is a safety component.

Picture 1 Semi-trailing control arm for rear axle, component parts

The <u>previous concept</u> was one assembly line each for the left and right control arm. This conventional assembly line with various mechanical assembly units to ease the manual operations is illustrated in picture 2.

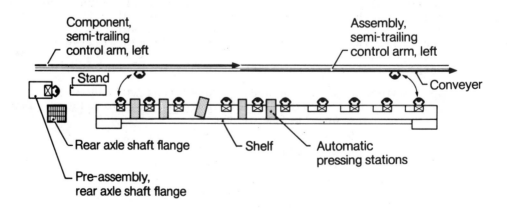

Picture 2 Assembly of semi-trailing control arm for rear axle on assembly line, using automatic assembly units

On this line it was hardly possible for the individual worker to vary his working rate. The assembly rhythm was determined by the machine. The job content was limited because of the short cycle of 7/10ths of a minute. Moreover, communication among the workers was severely restricted by the assembly machines between them (picture 3).

Picture 3 Semi-trailing control arm for rear axle, assembly-line production

On the one hand to counteract monotony by job rotation, but also for the sake of greater flexibility in the event of absence of a worker, all assembly workers were trained at all assembly stations with the exception of rear axle play adjustment and inspection. It was intended to practice job rotation, but in fact hardly anyone took advantage of this opportunity. The <u>new assembly system</u> (picture 4), with some 20 % greater capacity, is distinguished by the fact that the manual section is completely separated from the machine section. Thus the workers are disengaged from the working cycle of the machine section.

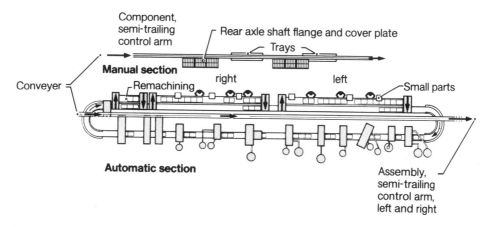

Picture 4 Assembly of semi-trailing control arm for rear axle in working groups

Although the working cycle in the automatic section of the new system has been shortened to one quarter of a minute, the job content for each worker in the manual section could be raised to 1,5 minutes, that means to a cycle which was agreed upon by the collective-bargaining parties for the area of these assembly workers as the desirable minimum working cycle applying to new manufacturing equipment.

Each worker takes the control arms and rear axle shaft flanges delivered by conveyer from the machining sections and after visual checking positions the parts on one of the assembly pallets in front of him (picture 5). He also installs the cover plate and brake carrier and attaches the fastening bolts. These workers are responsible for completeness and faultless from defects of the parts on each assembly pallet.

Picture 5 Semi-trailing control arm for rear axle, group assembly

At the end of the manual section the pallet enters the conveyer system of the automatic line section. This conveyer system also functions partly as a buffer line.

Other workers have to fill the magazines with different small parts which are then conveyed to the assembly stations by individual systems. These workers are responsible for the section they are feeding.

The previous method of adjusting the bearing play, which required a lot of skill, strength, and attention has been replaced in the new assembly line by an electro-hydraulic adjusting device and electronic inspection equipment. The worker only has the function of monitoring the unit and of selecting samples for random quality control.

Finally, the completely assembled control arm is hoisted from the assembly pallet and transferred to a conveyer to the final axle assembly department.

Despite the partial disengagement of the worker from the rigorous working rhythm of the assembly machine, economic operation of this assembly section could be achieved.

One important reason for this, however, is that now only one of the capital-intensive automatic sections with its feeding and testing equipment had to be purchased, since now both, left and right control arms for the rear axles can be assembled on this unit.

In addition, the danger of accidents, which was greater before when the workers worked between the machines, was reduced. The possibility of communication among the workers is fully guaranteed.

Unpleasant manual jobs such as applying sealing materials or lubricants - dirty and smelly jobs - were automated and hard physical labour was reduced.

3.2 Stub axle assembly

Another example is the assembly of the front stub axle, which includes attaching the front wheel hub to the brake disc and assembling these with the steering knuckle, brake cover plate, and dust cap (picture 6).

Picture 6 Stub axle, component parts

For the planning of the new stub axle assembly line the following requirements were to be met:

- creation of working places indepedent of fixed work cycles
- increase and enrichment of job content
- avoiding of hard manual labour
- increased flexibility of assembly operation
- increase of production capacity and economical operation

The operations for the first subassembly (wheel hub - brake disc) are limited to filling the magazines with wheel hubs, brake discs and small parts, monitoring the machines and equipment, random-checking the assemblies, and any necessary refinishing work.

The new assembly system effected a considerable reduction of the work load for tightening of the five brake disc fixing bolts. On the conventional assembly line the bolts had to be tightened to a torque of 11,5 mkp and marked. In the new system the initial and final tightening and marking are done by means of an accurate power torque wrench. Even the manual transfer of the relatively heavy wheel hub - brake disc-assembly to an overhead conveyer was automated. The overhead conveyer carries the wheel hub with the brake disc attached via lowering stations to the newly designed twin working stations.

A considerable portion of the previous assembly line work in the stub axle assembly was taken over by the total of seven twin working stations. Moreover, a possibly necessary increase in capacity can be effected by adding more working groups to the system (picture 7).

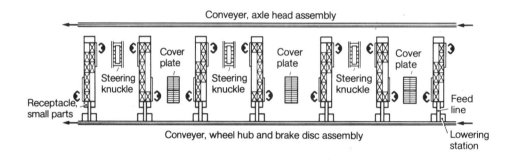

Picture 7 Stub axle assembly in working groups

The twin working stations are furnished with a rectangular pallet transport system Following the manual assembly, each pallet is released by the operator for transport to the next station (picture 8).

Picture 8

Stub axle - twin working place

Compared with the previous assembly line work, the job content was more than
doubled. Moreover, the work can be done at the new twin working stations without
stand-bys. A buffer line holding about 4 assemblies allows working ahead 7 or 8
minutes or slowing down. Each worker takes the responsibility for the correct
assembly of this important safety component, which he certifies with his inspection mark.

This assembly method of the front stub axle provides a solution which, favoured
by the reduced capital expenditure for the assembly equipment, has facilitated
a farreaching independence for the worker, along with expanding the job content,
lessening the manual energy required and the loads to be carried.

3.3 Differential gear assembly

Our last example is the differential gear assembly section put into operation
at the beginning of 1977. This assembly facility replaced the previous conveyer-
line assembly for our high-volume production line. The assembly contains the
differential parts - as shown in picture 9.

Picture 9   Differential gear, component parts

The manual individual work stations are no longer arranged in line, but
parallel. Each worker can set his own working rhythm, completely independent of
the mechanical section and the neighbouring manual working places (picture 10).

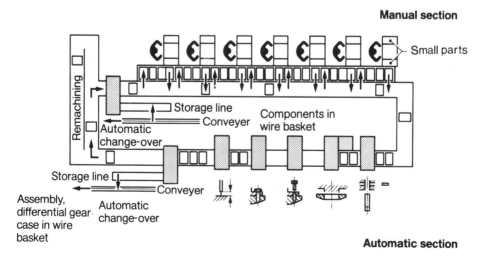

Picture 10   Differential gear assembly with individual working
             places arranged in parallel

4. OTHER SUBJECTS

AUTOMATION OF WIG-WELDING*

U. Lübbert

Institut für Informationsverarbeitung in Technik und Biologie, Karlsruhe, Federal Republic of Germany

Abstract. The observation of the welding spot of WIG-Welding on a television monitor is achieved by using a periodically intermitting source of welding current and by a television camera with a synchronous light chopper. The size of the welding loop is continuously measured by processing the videa signal in a simple electronic decive; in this way a full automation of WIG-Welding is intended.

INTRODUCTION

In welding technology there are two reasons for supervision of the welding process by means of technical sensors. The first reason is associated with the continuing development of automation; the other one is the desirability of using industrial robots for welding at difficult or dangerous locations (Becken 1970). There are not only spectacular applications, such as welding within the radioactive zone of a nuclear reactor, but also more down-to-earth ones, such as in pipeline and steel constructions. For instance, in manufacturing of large steel structures, welders have to work inside boxes preheated to 80°C, in physically stressful postures.

ADVANTAGES OF OPTICAL SENSORS IN WELDING TECHNOLOGY

Simple mechanical sensors have frequently been used in many fields of application, e.g. in mining technology. In a number of cases they have been replaced by light beams or by other optical arrangements. However, these sensors only enable a binary output, or the display of a scalar measuring variable. Contacting sensors are not suitable for the welding process in view of high temperatures. Obtaining scalar quantities has not been successful, such as the temperature or the the degree of material smelting in individual locations. For these reasons, the automation of welding operations has not yet been augmented above the level of feedforward control.

With this type of control all relevant parameters have to be kept sufficiently constant; hence one has not yet been able to solve the problem of automatic rewelding of "Heftstellen bei Wurzenähten". Sensors for scalar measured variables are not suitable because the information about process operation is distributed over many conditions. For welding under inert atmospheres the information is mainly to be found in the picture of the welding spot, hence a picture detector, e.g. a television camera, is a necessary component of a sensor for welding technology.

The purpose of an optical sensor is picture observation and processing; so pictures will be evaluated according to predetermined points of view, and consecutively the desired operation will be realised, depending on contents of the picture. Consequently an optical sensor for arc-welding shows the following functions:
Optical part for projecting the observed process on a photoelectric layer;
Optical-electronic transducer for transforming optical signals into dectrical signals;
Data processing part for extraction of characteristics important for welding and for transformation to an output signal of the welding process.

THE USE OF A TELEVISION CAMERA (Lübbert)

Correlation techniques have been applied to the automatic recognition of optical samples, where the correlation between the picture to be recognised and reference pictures is of significance (Muhlenfeld 1969). The two-dimensional correlation is being realised in an optical way, partly with coherent and partly with incoherent light. A sensor for welding can only process the incoherent light

---
*) In view of time, this paper has been translated from German into Engli'sh by J.E. Rijnsdorp. He should be held responsible for any errors and inaccuracies.

emmitted by the welding spot. An optical correlation, for instance with optical masks, has proven to be insufficient for data-processing of welding pools, irrespective of its relevant simplicity. Comparison with fixed masks is insufficient due to the varying width of the welding seam and the movements of the arc, moreover it is impossible to evaluate structures with little contrast, which are at the border between solid material and melt.

In contrast, the application of a television camera offers considerable advantages. Picture decomposition into a time-dependent luminance signal enables electronic preprocessing, e.g. the augmentation of structures with sharp edges by taking the derivative, and the augmentation of structures with a certain luminance by non-linear threshold operations. Moreover, since picture preprocessing has resulted in a time-dependent function, it can not only be subjected to the same operation as can be realised by an optical correlator, but it also offers the possibility to make processing invariant with respect to certain picture distortions.

The application of a television camera is not without problems because of the high intensity of the electric arc; it only became possible through the development of a Vidicon with a silicium-diode-target, which has high sensitivity in the infra-red and is insensitive to overexposures. Only extreme thermal influences can destroy of "burn-in" the picture. Its price is lower than that of a Plumbicon.

The picture of the welding pool contains strong differences in intensity between the welding pool proper and the electric arc, hence the dynamic range of a television camera is insufficient (Becken 1966). In order to circumvent the extinction of the picture of the welding pool by that of the electric arc a method has been developed in the context of this work, which allows television transmission of the inert gas welding process with high quality for the first time. Here the experience has been applied, obtained with pulsed welding current sources.

The welding current is periodically (50 Hz) strongly attenueted or switched off during short intervals of time. In these intervals the welding pool radiates without the disturbing influence from the electric arc. The optical path from welding location to vidicon is only opened during these time intervals by means of a synchronous light chopper. Although the vidicon is only illuminated during brief time intervals, this is sufficient for generating a charge picture which can be scanned in the usual way. The picture quality is so good that details of the welding process can more comfortably be observed than directly through a welding filter.

THE APPLICATION TO WIG-WELDING

The taking and processing of the picture of the welding spot is now applied to the control of a simple welding process. As an example WIG (Wolfram Inert Gas) Welding has been chosen.

In this inert gas method the electric arc is between a non-melting tungsten electrode and the work piece. The addition material is introduced sidewise. Argon is brought to the welding spot through an annular nozzle around the tungsten electrode in order to avoid oxydation. In the trials steel sheets have been welded in a 60° V-seam. These sheets were made of steel 37, and had edges cut by fire torch. The opening was fixed at a width of 2-3 mm.

The investigations were started in the Welding School and Laboratory Duisburg; possibilities were investigated for step-wise automation of the welding process, starting from manual welding. Then the WIG-burner was made to progress with constant speed, subject to a constant swinging motion corresponding to the movements for manual operation. Here the welder only had the possibility to feed the addition material. In the second step, the addition material was supplied through a nozzle, hence it could only be immersed into or removed from the welding pool in axial direction. In the third step, also this movement was realised by a special mechanism, which could be operated by a switch.

Under these conditions the mechanical arrangement is suited for control by a single correcting condition, the intermittent supply of addition material. Man participates in the control loop by operating the switch for supplying the addition material; he observes the form of the welding spot, which represents the controlled variable.

The process starts with melting the end of the corner location. The edges of the working pieces are being melted by means of the swinging progression of the burner. The welding loop is formed, an annular extension of the opening (Fig. 1):

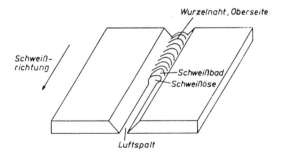

Fig. 1: Partly welded sheets, showing opening, welding pool and the top of the weld.

The welder manipulates the switch as soon as the welding eye has passed a certain size. This introduces the addition material into the welding pool, a well-defined piece is melted, and the addition material is withdrawn in order to observe the diminution of the size of the welding loop. Then the welding loop is again augmented in size due to further progress of the burner, hence the addition material is supplied again at the correct moment of time.

With the described set-up also unskilled persons are able to generate straight welds of high quality. They were asked to introduce the addition material exactly when the welding loop passes a certain size as seen on the television monitor. In this way it was shown that an excellent weld results from the reduction of the number of degrees of freedom to the moment of supplying the addition material and that measurement of the size of the welding loop is sufficient for process control. This result does not hold for skilled welders because they subconsciously watch for other criteria.

ELECTRONIC PROCESSING OF THE WELDING LOOP PICTURE

In order to verify the results a set of welding loop pictures were investigated with respect to size and shape. Figure 2 shows three different welding loops; too large, optimal, and too small.

Fig. 2: Different welding loops, direction of welding is downward.

Fig. 3 shows the contour of a welding loop in a schematic way. Fig. 4a shows a series of different types of welding loop which have been sketched from photographs.

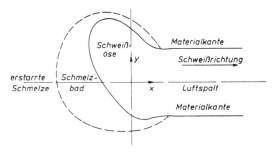

Fig. 3: Contour of a welding loop (schematically).

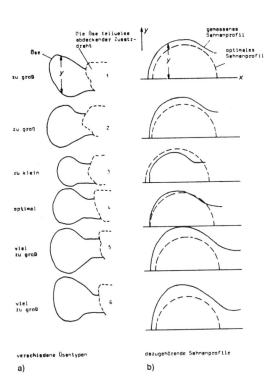

Fig. 4: WIG-Welding
a. Different types of welding loop
b. corresponding chord-profiles.

The surface tension minimises the area of the welding pool; it can therefore be expected that the basic shape of the welding loop approaches a circle, if the heat introduction and the blowing effect of the electric arc are completely symmetrical. Figure 5 shows such an idealised welding loop.

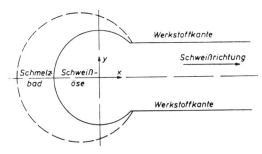

Fig. 5: Idealised welding loop.

When the circular shape is distorted due to an asymmetry in the Y-direction (perpendicular to the welding direction), one can presume that the chord-profile (the chord-length in the Y-direction as a function of location X in the direction of welding) remains almost invariant. Fig. 4b shows the chord-profiles corresponding to the welding loops of Fig. 4a. Welding loops 1-6 have analogous chord-profiles, which corresponds approximately to the ideal circular shape.

The circle diameter, corresponding to the chord profile is thereby the significant criterium for supplying addition material. This is illustrated by the dashed semicircles in Figure 4b, corresponding to the profile of loop 4, which is considered to be optimal. The welding pool can still be maintained with all these welding loops which indicates the sensitivity of the criterium.

The welding loop (Fig. 3) is scanned according to the TV-standard line by line in the Y-direction by a silicon-diode-vidicon. The videa-signal from the television-camera represents the picture luminance along a line. Figure 6 shows the pattern. In the video signal there are pulse shaped depressions,

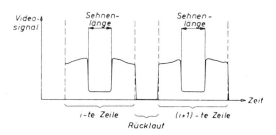

Fig. 6: Video signal from scanning the welding eye.

with a duration proportional to the corresponding chord-length. An especially developed derivative circuit reliably extracts the welding loop contour from the remaining picture, also when there is a varying background illumination and a small residual electric arc. The chord profile is being measured, and the corresponding circle diameter calculated by means of simple digital circuits. The system can be adapted in a flexible way to the pertaining weld opening by changing units of measure, which can simply be adjusted by electronic means.

FUTURE

The optical sensor described in this paper enables observing the welding process from a distance, hence it contributes to humanisation of the welder's work location. Present investigations have the purpose to realise control of WIG-Welding by continuous measurement of welding-loop size by means of the indicated electronic processing circuit; hence full automation can be obtained. Extension to other inert gas welding methods seems promising.

REFERENCES

Becken, O. (1970). Problematik der Entwicklung von Schweissrobotern, Schweissen und Schneiden. Band 22. S. 2 - 4.
Lübbert, U. Deutsche Patentanmeldung Nr. P 25 00 182.5.
Mühlenfeld, E. (1969). Eine Einführung in die optische Informationsverarbeitung, ISF-Mitteilungen, S. 19 - 23.
Becken, O. (1966). Versuch einer Fernsehübertragung des Schweissvorganges, Schweissen und Schneiden 18. S. 60 - 63.

SEARCH OF AN APPROPRIATE TRANSFER OF TECHNOLOGY TO A DEVELOPING REGION WITHIN AN INDUSTRIALIZED COUNTRY

Aldo Dell'Oro*, Umberto Pellegrini** and Claudio Roveda***

*University of Calabria
**State University of Milano
***Politecnico di Milano

1. Introduction

The existence of different rates of economic growth among different countries and among different regions in the same country has been a common characteristic of process of economic development (1). Recently the phenomenon has become even more acute and critical: the exceptional rise of the price of oil in 1973 has had a different impact on the countries of the so-called "Third World" and has given rise to a new class of countries, the "Fourth World" i.e. underdeveloped countries with no oil resources, which are going to see their economic conditions worsened. The situation of uneven development is more apparent among countries: therefore a large amount of studies on the conditions of economic growth has been undertaken by economists in the last two decades.

But also within a country, especially an industrialized one, inequality of level of development among its regions has become more acute and widespread, with remarkable social consequences. The concentration of economic activities in certain regions and the territorial and sectorial unbalance which follows, represent a cost for the whole country, as its regions and industrial sectors do not give their full potential contribution to national product. Hence the necessity of finding ways to activate fast development processes in underdeveloped regions.

Quite recently, it has been recognized that technological innovation, and transfer of technology, can be a very useful tool for economic growth of underdeveloped countries. This attention to the problem of transfer of technology has become widespread among both scientists and politicians, together with the awareness that it is necessary to adapt technologies to the conditions (cultural, social, environmental, etc.) of underdeveloped countries, in order to start effective processes of growth.

In this paper we takle the specific problem of transfer of technology to underdeveloped regions of an industrialized country (in particular, the Mezzogiorno of Italy). After evaluating which rôle, and under what conditions, transfer of technology can play for economic growth in such a case, we outline a methodology for defining the lines of an appropriate industrial and technological policy (*).

2. Appropriate technology

Past experience of international transfer of technology through establishment of production subsidiaries of multinational companies, creation of joint-ventures between local and foreign companies, licensing, activity of international organizations, etc., has taught that it is not enough to transfer hardware, but that it is almost necessary, in order to have real success, to develop appropriate software, i.e. know-how, technical capabilities, organizational structures, etc. (2). In other words, the technology to be transferred must be "appropriate" to the local conditions and to the objectives of socio-economic growth of the region, where usually local capital is scarce, man-power is over-abundant and poorly trained, entrepreneurs are few and not very active, infrastructures and external economies are missing.

"Appropriate" technology has almost always been considered as synonymous of intermediate, low-cost, labour-intensive, small-scale, soft technology.

In a Seminar organized by OCDE in 1974, F. Stewart (3) listed the following nine characteristics of appropriate technology:
- low level of fixed capital per unit of man-power
- low level of fixed capital per unit of product
- low unit cost of machinery

(*) The main results of this paper have been derived from a larger investigation on a new industrialization model for Mezzogiorno, which is presented in : FAST-CSATA "Trasferimenti tecnologici e sviluppo industriale del Mezzogiorno"
Milano - September 1977

- simplicity of production, maintenance, management processes
- connection with the traditional sector of economy
- connection with agricolture, which is the predominant economic activity in underdeveloped countries
- small-scale plant
- use of local raw materials and resources
- low requirement of foreign financial and technical resources.

The correlation between these characteristics and the problems of industrial development of underdeveloped regions is quite apparent: but it's also clear that such definition of appropriate technologies is biased towards the problems of the Third World, and cannot be used as it is, for a region such as Mezzogiorno. This region is certainly underdeveloped, but it is also part of an industrialized country, which is an open economic system, fully integrated in the international market. As it is not possible for Italy to follow protectionistic and autarchic economical policies, it follows that all industrial ventures in the Mezzogiorno must be competitive, at a certain degree on international level.

More than this, there are two important factors which strongly differentiate the situation of Mezzogiorno from underdeveloped countries. The first is the cost of man-power, which is in the Mezzogiorno almost at the same level of all other industrialized countries and is much higher than in underdeveloped countries. The second factor is the quality of man-power: in the Mezzogiorno, unemployed people are, mostly, young people with a good level of education.

Any industrial policy for the Mezzogiorno must take into account these constraints and resources: international competitiveness, high cost of labour, young and educated potential working force.

Therefore, the technologies appropriate for the Mezzogiorno must be labour-intensive, have small-size economical plants, be raw material- and energy-saving, avoid environmental pollution and congestion; they must have characteristics opposite to the ones of the model of industrialization that has been followed up-to date, that is, they must be "soft" technologies, rather than "hard" technologies (4). In any case, they cannot be intermediate technologies, but they must allow a high productivity of labour, without using a high level of fixed capital, that is, must be "advanced" technologies.

To this aim, they must make the largest and fullest use of the scientific and technical knowledge that has been produced up to now in every country and is available from the scientific community.

At this point, we can propose a different definition of appropriate technology, as a technology characterized by:
i) capability of using large amounts of man-power, either direct or indirect
ii) accessible know-how
iii) not difficult management and organization of production process
iv) low level of fixed capital per unit of man power
v) economic feasibility of small-scale plants.

3. Outline of a methodology for establishing priority among industrial sectors on the basis of appropriateness

The selection of appropriate technologies gives rise to two different problems:
1) given a certain industrial sector, find the "appropriate" technology among the various methods of production (existing or to be developed).
2) at macroeconomic level, find the sectors which are more appropriate.

In the first case, the choice is among different technologies for the same product; in the second case among different sectors, which use technologies with a different degree of appropriateness. Of course, in the first case technology is used in the traditional sense, as scientific and technical knowledge applied for the production of a certain item; in the second case, technology is used in a broader sense, as a set of homogeneous processes for a class of similar products.

The development of appropriate technologies in a given sector requires new directions for R&D, both at academic and industrial level: i.e. in many manufacturing sectors, it is needed to develop energy-saving systems; in the chemical sector, to find new anti-parassites products, which do not alter the biological equilibrium, etc.

On the other hand, the establishment of a hierarchy among sectors requires both a more specific and quantitative definition of appropriateness and a theoretical and empirical analysis of technological features of the industrial sectors.

This paper takles the second problem and presents a methodology for the choice of appropriate sectors: the methodology has been developed for the specific problem of the Mezzogiorno of Italy, but it is quite possible to extend its use to the case of developing areas in other industrialized countries, given the similarity of conditions and situations.

The starting point of the scheme is the

analysis of the balance of trade for the various sectors, in order to find out those for which national demand is not fully satisfied by national production: in these cases there exists a national market and there is the possibility of expanding national industry. This approach aims at developing import-saving production and, of course, it is not the only possible one: anyway, it is important because it tries to overcome the constraint that deficit in the balance of trade impose on the economic development of a country.

The second step consists of the evaluation of the parameters which define the degree of appropriateness for the sectors which have a strong deficit of the balance of trade and for which imports are a high percentage of internal demand. This evaluation is usually rather complex and requires the collection of large amounts of data.

The third step consists in referring the results of the previous purely economic analysis to the actual conditions of Mezzogiorno (or developing region), in order to find out the feasibility of the location of the various sectors in the region and define the necessary actions on infrastructures, education, etc.

The efficiency of transfer of technology depends heavily on the creation of appropriate conditions in terms of generalized external economies in the Mezzogiorno: and this is a specific task of the State and public administration.

We can now analyze the three steps of the proposed methodology with some more details.

3.1. Analysis of the balance of trade

Preliminary to any analysis is the definition of industrial sector.

Whatever criteria are chosen, subdivision into sectors must be very detailed in order to avoid use of aggregate data, which might average quite different situations. A limit to a detailed subdivision comes from the availability of appropriate raw data: in Italy official statistics are usually quite aggregate and more detailed data can be obtained often only through specific and direct investigations.

The definition of sector can be done in many ways, at least in theory, the most common ones are: 1) aggregation of production activities of very similar products (intermediate or final); 2) aggregation of all production activities of items which are a part of a given final product (vertically integrated sector). Both definitions have advantages and drawbacks. The first one is useful because it allows a specific analysis of technological characteristics, but it might give rise to a too narrow point of view so that it is difficult to give meaningful proposals of intervention.

On the other hand, the vertically integrated sector has the advantage of specifically taking into consideration the interrelations among industries, so that it is easy to evaluate the national added value and the rôle of national firms in the overall production cycle. The drawback is the difficulty of a technological analysis, given the very different characteristics of the production processes involved in the cycle.

A third empirical criterion, in some case quite useful, is to assume the same subdivision as that of business associations.

After assuming one of these definitions or a mixture of them, the task of this step is the analysis of the balance of trade, in order to find out sectors with a high deficit either of final or of intermediate products. The parameters to be evaluated can be:
i) absolute value of deficit
ii) trend in this deficit for the period 1973/1976, given the special rôle of 1973 in international economy
iii) ratio of imports to internal level of demand
iv) ratio of imports of the sector to total imports of industry.

From these parameters one should be able to derive a simple indicator to be used for ranking the sectors in order of relative opportunity of development. It seems quite difficult to find a precise and rational mechanism for such indicator. Anyway, one should combine quantitative with qualitative analysis: the data on the balance of trade must be interpreted on the basis of the various factors (technological, entrepreneurial, organizational, political, etc.) which influence the level of international competitiveness of the firms of the sector (*).

3.2. Building of a matrix "Sector/Appropriateness"

For each sector chosen in the previous step it is now necessary to evaluate its appropriateness through some quantitative parameters, from which one can derive a single indicator of appropriateness.

In this way, a matrix "Sector/Appropriateness" is built and is the basis for the choice of the sectors which, in theory, are susceptible to be used to start an economic growth in the Mezzogiorno.

(*) An example of this approach is given in the research project carried out by these authors, sponsored by the Minister of Foreign Trade of Italy, on the relationship between flows of trade and quality of Italian firms in some sectors.

We say in theory because the characterization of the sector is done in purely technological and economical terms, without taking into account the specific urban and environmental conditions of the Mezzogiorno: this analysis will be done in the third step.

Following the definition of appropriate technology previously given, we propose to assume as its parameters: i) the value of fixed capital per unit of direct manpower; ii) the value of fixed capital per unit of product; iii) minimal economic scale of production (in terms of number of workers).

The measurement of other characteristics of appropriateness (impact on environment, organization and management of the production process, etc.) is quite difficult: one has to rely on qualitative evaluations to be gathered from experts, scientific and business associations and trade unions, if any.

The overall indicator of appropriateness can be evaluated as a weighed average of the various parameters, whose weights are to be determined in a qualitative way.

The choice of the sectors has to be based both on the characteristics of the balance of trade and on the level of appropriateness: in case these parameters are conflicting, one can use well-known procedures for the selection of alternatives with multiple criteria (5).

3.3. Feasibility analysis of location in the Mezzogiorno

This step starts with the analysis of the economic and industrial structure, in terms of sectors, occupation, output, etc., both in absolute and with reference to the whole of Italy. In this way, it is possible to evaluate how the proposed sectors can match, and develop from, the existing economic structure: this evaluation cannot be carried out in a completely deductive way. One has to rely on information and analyses carried out by business, scientific and technical associations and experts, and to put them together into an overall rational and coherent framework.

As to the location factors, there are available many theoretical and empirical studies: the most useful of them (6) presents an interesting matrix, where for each kind of industrial plant (broadly defined) a set of location requisites are evaluated (technical training of man power, consumption of water, infrastructures for transportation, etc.) and from them an overall indicator is calculated, as a weighed average. These results, or other similar, allow to complete the analysis of the feasibility of developing the previously defined appropriate sectors.

4. Concluding remarks

The here proposed method does not pretend to be an absolutely rational tool to define the industrialization policy of the Mezzogiorno: it gives, anyway, a rational basis for the choice of the sectors to develop, choice which has to be made at political level.

Even if the ranking of the appropriate sectors has not an absolute value, given the subjective values of the weights to be used, it is possible to define a tipology of sectors in hierarchical order:
1) sectors, with trade deficit, high degree of appropriateness and without constraints for location
2) sectors, with trade deficit, high degree of appropriateness and with some constraints for location
3) sectors, with trade deficit, low degree of appropriateness, with or without constraints for location.

Another set of problems arise for the implementation of the chosen industrial sectorial policy. It is apparent that one cannot rely on the cultural, scientific and entrepreneurial forces of the Mezzogiorno. The appropriate technology cannot be generated by these forces, but must be transferred from outside: in a short and medium range, there is no need of new laboratories and new activity of R&D.

It is necessary to make use of all the scientific and technical know-how which has been produced in the world and to transfer this know-how into operational industrial activities.

Without trying to cover the subject of how to make the required transfer of technology (7), from one investigation there emerge some definite points:
- transfer of technology cannot be done only by forwarding technical information to entrepreneurs
- transfer of technology requires the use of all scientific and technical forces of the country
- technical information, to be searched and acquired at international level, must be transferred into industrial operations by using all these forces
- new institutions, specifically aimed at transfer of technology, must be set up, to put a bridge between the scientific and the industrial world.

References

(1) G. Myrdal, Economic Theory and Underdeveloped Regions, Duck Worth, London, 1957.

(2) N. Jéquier (ed), Appropriate Technology: Problems and Promises, OCDE, Paris, 1976.

(3) F. Stewart, Intermediate technology, a definitional discussion, in Choice and Adaptation of Technology in Developing Countries, OCDE, Paris, 1976.

(4) E.F. Schumacher, Small is Beautiful. A Study of Economics as if People Mattered, Blond and Briggs, London, 1973.

(5) B. Roy, Problems and methods with multiple objective function, Mathematical Programming. 2, 239-66 (1971).

(6) Centro-Piani, Tipologia industriale ed infrastruttura del territorio per una politica di sviluppo del Mezzogiorno, Paper 15, Roma, 1976.

(7) D. Gurmidis, Le transfert de technologie, Tiers-Monde, XVII - 65 (1976).

AN APPROACH TO THE PRODUCTION LINE OF AUTOMOBILES BY MAN-COMPUTER SYSTEM

K. Namiki*, H. Koga*, S. Aida** and N. Honda**

*Honda Engineering Co. Ltd, Sayama-shi, Saitama 350-13, Japan
**University of Electro-Communications, Chofu-shi, Tokyo 182, Japan

Abstract. In order to improve the manufacturing capacity of a production works substantially, mere improvements of performance in "hardware" such as process machinary and upgrading of production workers are not sufficient, and a fully integrated total production control system closely adapted to the actual conditions at the production site must be formed. It is necessary to develop a method, based on the concepts of system engineering, of analysis and synthesis for such problems of the production works. In this paper, two methods, formed in man-computer system using the functions of man and computer, are proposed. Each effectiveness of two methods is discussed in relation to examples taken from the problems of line balancing and stock minimum, respectively, of the assembly line in the production of motor cycles or automobiles.

Keywords. Optimal search techniques; Man-machine system; Manufacturing process; Computer software; Feedback.

1. INTRODUCTION

The assembly line in the production of motor cycles or automobiles is one of the labor-intensive job sites where the greater part of the works still relies on manual labor, although various attempts have been made to develop automated systems. Roughly speaking, there are two types of steps for regaining the dignity of man, which is considered to have been lost in the job site known as the assembly line. One of them tries to improve the situation by automating all types of works that are not suited for human labor. The other, transient in nature as it may seem, tries to regain it by making working conditions in the assembly line more suitable for human labor. As far as automation in the assembly line remains at its current standard, it is clear that automation is inferior to manual labor in attaining economy and flexibility. In order to raise the standard of automation in the assembly line. further studies should be made over a long period of time. On the other hand, the latter or making working conditions in the assembly line more suitable for human labor is more practical as a tangible step, for which a methodology is discussed in this paper.

A system can be modeled into a structure on the basis of a set of elements together with a set of relations formed through a definition among their elements. The system structure of an assembly process can be formed on the basis of the set of minimum rational work elements (MRWE) and the precedence relation among the MRWEs. Recently, in the field of studies of social systems, some methods are being developed for the structural modeling of complicated systems. Interpretive Structural Modeling (ISM) as one of the above mentioned modeling methods is provided with a function capable of structural modeling of complicated problems through the repetition of conversation between man and computer. An assembly process is a system provided with the structure which is capable of satisfying various with regard to time and space, based on the precedence relation among the MRWEs.

Some conventional methods for the problems of an assembly process have the limit of their effectiveness because of depending on only computer function. Our developed methods, based on the cooperation of man and computer functions, are shown in the following.

2. THE PROBLEMS OF LINE BALANCING IN ASSEMBLY PROCESS

2.1 Line balancing

The purpose of line balancing is to distribute to workers the load of the MRWEs in an assembly process as equally as possible, according to the precedence relation among the MRWEs, when a number of workers are positioned along a conveyor for assembling one kind of product on a flow production.

Recently, a number of studies of line balancing programs have been conducted, but none of them has been practically applied. The reason for this is that the problems of line balancing are settled only with computer. The problems of line balacing are summarized

as follows:
1) There are many points which can hardly be formulated on a logical basis, because operation in any assembly process can be maintained with the help of men --- ambiguity.
2) Line balancing is a combination problem, and there is no effective way available to obtain a practical and perfect solution about this problem --- complicity.
3) The purpose of line balancing is to exercise control over men, but it must be men themselves who make the decision regarding such a matter. Men must be not controlled by an assembly process --- ethicality.

2.2 Approach to Line Balancing

As seen in the above, the problems of line balancing should be handled by man, but these problems could be handled more effectively when supplemented by computer. The best way to handle these problems is to clarify the boundary between the work that can be done only by man or that is suited for man, and the work that is suited for computer. In other words, the best way is to harmonize the ability in which man is superior to computer, such as pattern recognition, learning, multi-dimensional evaluation and decision making, with that in which computer is superior to man, such as information processing. Table 1 shows the allotment of works between man and computer in line balancing.

In order to formulate a line balancing system in cooperation between man and computer through conversation, the problems of line balancing can be roughly divided into four sub-problems.
1) Structural modeling of the work order in an assembly process. To prepare a set of the MRWEs before making a structural graph of an assembly process, according to the precedence relation that is to be defined on the set of the MRWEs.
2) To set up various restricting conditions and to determine the weight of evaluated items.
3) To handle combination problems, paying attention to line balancing.
4) To evaluate and modify the results of line balancing.

Fig.1 shows the formation of the problems of line balancing composed of the four sub-problems. In order to solve these problems, a structural modeling process through conversation between man and computer is required and a methodology, based on the concepts of ISM, is considered. Such a methodology is named Interpretive Structural Modeling of Assembly Processes (ISMAP), the general diagram being shown in Fig.2. As distinct from ISM, which is capable of structural modeling only on a qualitative basis, ISMAP is capable of that in any assembly line not only on a qualitative basis but also on physical as well as quantitative basis.

3. INTERPRITIVE STRUCTURAL MODELLING OF ASSEMBLY PROCESSES (ISMAP)

3.1 The Contents of ISMAP

Fig.3 shows a process in which ISMAP is applied to the formation of a practical line. First of all, an assembly process as an object is determined and the MRWEs in it are extracted on the basis of design data on products, assembly methods, etc. When the extraction of a set of the MRWEs is finished, the precedence relation that is to be defined on the set of MRWEs is extracted. Fundamentally, this process is to be performed through pair comparisons between two MRWEs, repeatedly, Experts are expected to answer a series of questions such as "Does MRWE 'X' come before MRWE 'Y'?", while the ISM program is used for ranking these works, based on the answers, according to the transitivity among them. With this process repeated, the qualitative structure of the assembly work can be clarified. The diagram in the precedence relation thus prepared will be used as the input data for the line balancing program. With the processes mentioned above, the assembly work is modeled into a structure on a qualitative basis. However, the assembly line can be formed only after the line is examined with regard to time and space conditions, based on the structural model of the assembly work thus obtained. The line balancing program is mainly used for dealing with this process. To this end, it is necessary, in advance, to set up restricting conditions and to determine the weight with regard to the evaluation functions. Cycle time is usually determined in accordance with the level of the most important production plan, and the upper limit of the rate of operation is determined with due consideration to human job conditions. Then, restricting conditions with regard to space, such as working position, and the characteristics of each MRWE are to determined. The diagram of the precedence relation and various restriction conditions thus determined are used for preparing the input data for the line balancing program, and a substitutional plan for line balancing is made with the aid of computer. Hereupon, the person in charge of line balancing evaluates the results thus obtained with due consideration to various points with regard to time, space, human engineering and physical points before modifying them, by comparison of the results with an ideal line formation, with which he is acquainted, in the assembly line. An assembly line formation plan for an almost ideal line formation will then be completed after all these processes mentioned above have been repeated several times.

3.2 The Application of ISMAP

The line described here is an assembly line for the frames of motor cycles, which is relatively simple in formation. The characteristics of the line and the work are given as follows:

1) This is a flow-shop type assembly line composed of 27 MRWEs (a single model line).
2) In order to carry out the assembly work, workers are positioned either to the left or the right side of a belt conveyor.
3) Parts needed at each process are supplied either from the left or the right side of the line.
4) The number of workers positioned at the existing line is in approximately 17.
5) Cycle time is in the range from 16 to 19 seconds.
6) The upper limit of the rate of operation for the workers is 0.8.

A set of the MRWEs is prepared on the basis of the actual assembly process. Table 2 shows working hours, the number of workers required and the working position. In order to carry out a structural modeling basing on the precedence relation in the set of these MRWEs, the ISM program was used for processing the data matrix while experts on the job site were asked to remodify the results thus obtained, and this process was repeated several times to prepare the diagram of the precedence relation as shown in Fig.4.

The restricting conditions to be set up and the weight of the evaluation functions are determined as follows:
(Restricting conditions)
The following setting was made with
   The interval of cycle time...17 to 19 seconds
   The upper limit of the rate of operation.................0.8
   Positional restrictions......matrix of moving time as follows:

	R	L	B
R	0	∞	0
L	∞	0	0
B	—	—	0

(Note) Movement of B→R and B→L depend on the position of workers.

(To determine the weight with regard to the evaluation functions)
The evaluated items:
   1. The rate of operation of workers
   2. Idle time
   3. Physical load

The items 1 and 2 are essential conditions for increasing working efficiency, while the item 3 is an essential condition for regaining the dignity of man, and these evaluation functions are defined as the followings:

1. The rate of operation Q_i = $\dfrac{\text{(The total of working hours alloted to process i)} + \text{(Working hours of } J_x \text{ work)}}{\text{(Cycle time)} \times \text{(The number of workers alloted to process i)}}$

2. Idle time t_{id} = (The hours taken for the J_x work of i process and the work alloted) − (Net working hours)

3. Physical load of worker S_p = (Physical load of workers who are alloted with the J_x work) − (Average physical load per a worker)

$$S_p = \begin{cases} 0 & (S_p < 0) \\ S_p & (S_p \geq 0) \end{cases}$$

Accordingly, the function for the evaluation of line balancing is defined:
   PI = $(1 - Q_i) + \alpha_1 t_{id} + \alpha_2 S_p$,
where α_1 and α_2 are determined by the designer of the line system.

We obtain the following results of line balancing by ISMAP. By keeping a cycle time of 18.0 seconds constant and by setting the rate of operation to the upper limit, the difference between one case in which an attempt is made to balance the physical load of workers and the other case in which no such attempt is made is shown in the following:
 a) Changes the rate of operation
    It is anticipated that the rate of operation decreases more greatly in one case in which an attempt is made to balance the physical load of workers than in the other case in which no such attempt is made. In other words, the more number of workers are required in the former than in the latter. However, it was possible to balance the physical load of workers without decreasing the rate of operation in this example.
 b) Changes of physical load balance
    Changes of physical load balance are shown in Fig.5 and 6. It can be seen with these graphs that a considerable improvement is noted in balancing physical load. In this case, the upper limit of physical load is not set. If it is set, a greater improvement can be expected in balancing physical load, but it involves a great risk of decreasing the rate of operation.

3.3 The Characteristics of ISMAP

The characteristics of ISMAP are summarized as follows:
1) ISMAP is an all-out structural modeling theory. In other words, it is a methology for the purpose of improving the production process through structural modeling.
2) On the theory of ISMAP, man himself takes the lead in carrying out line balancing with the aid of computer.
3) On the theory of ISMAP, an almost ideal line formation can be obtained, since man himself seeks a solution on a heuristic basis.
4) Simple algorithm is adopted for preparing the line balancing program, so that it is rapid in computation, and well suited for repeating processes. The language used in this method is FORTRAN with approximately 350 steps.
5) Learning efficiency for the persons in charge of line balancing is obtained.

4. PRODUCTION CONTROL SYSTEM AND INVENTORY

4.1 The Characteristics of Production System

A method which is for stock minimization in a production system is discussed in 4 and 5 chapters. Production system often comes into the system conflict between master production plans which are logically formulated by the top management (top-down plans), and the aggregate of demands which are human reactions to the change of production conditions and are directed from the production site upward to the top (bottom-up demands). This conflict is one of the most important factors preventing the improvement of efficiency in production. This method is made to successfully harmonize "top-down" plans with "bottom-up" demands by means of the cooperation of man and computer functions and represents a job shop scheduling system formulated to reduce the in-proxess inventory, and thereby to improve the efficiency of the production line and to free the related workers from the unnecessary working load.

4.2 The Problems of In-Process Inventory

The in-process inventory not only the safety, provided for such purposes as securing parts which are partly assembled by the other shop of the process and partly supplied by an outside maker, to avoid an eventual delivery delay, but also the lead stock, provided in preparetion for die changes. Especially, between two processes of different basic production forms, such as between a press shop treating with parts piece by peace and a welding shop assembling several parts together in the production line of an automobile body, a certain lead stock is provided of necessity. Although an in-process inventory is present almost everywhere in an production process, the stock between two job shops of different types such as represented by the press shop and the welding shop, will mostly be analysed. A production control system specifies "how much" of "what" product to be shipped out of the works by "when", as an absolute attainment goal. A production control system, also specifying the materials for parts and the delivery schedule for purchased parts, is a "top-down" directive, and is a "push" information system. On the other hand, as shown in Fig.7, a shop shipping schedule-(A) provides basis requirements for the process schedule of the preceding assembly shop-(B), and this process schedule-(B) in turn specifies the process schedule of the paint shop-(C). In this way, the flow of information in a production control system takes place in the direction opposite to the flow direction of objects (material parts semi-products products) in a production line. This means the process schedule of a succeding process is accepted as a set of restricting conditions in formulating the process schedule of a preceding process, and in this sense, we have a "pull" information system. In formulating the process schedule of the press shop-(E), the requirement that all the necessary parts must be supplied without any missing to the welding shop schdule-(D) is taken as the prerequisite. Press shop schedule means the determination of the lot sizes and the process sequence of the parts involved, specifying the precedence relation among various elements, on the basis of the capacities of the press machines and the workers, as well as the elements of production actividies such as the number of processes and the cycle times of the parts to be processed. The stock between the preceding press shop and the succeeding welding shop arise out of the time series correlation between the process schedules of the two shops. In this sense, the stock of this nature may be said to arise where a "top-down" directive and a "bottom-up" initiative interact.

4.3 Die Change and Stock

The die changes of press machines must suspend the production work for a certain time, reducing the production efficiency to some extent, and therfore raising the production cost correspondingly. On the other hand, if the lot size of various parts is increased to reduce the frequency of die changes, the lead stock these parts is enlarged and, because stock mean "dead" investment, there will be otherwise unnecessary interest cost added to the production cost. An optimum lot size must be determined on the jucicious evaluation of various contradiction factors concerning both the die change and inventory such as line efficiency and cost effectiveness. The stock size is subject to change with the thpe of process sequence, according to time series correlation to the welding schedule. With the ever increasing diversification and amount of products, the kind of parts has increased greatly, so that the problem of reducing the stock between two job shops to a minimum, based on correct grasp of the precedence relation among the various factors related to the lot size and process sequence in the production system, has become very important. This is not only important for the pursuit of cost merit, but also for the improvement of working efficiency and the load reduction of the workers in charge of the die changing operations and stock control.

5. THE METHOD OF STOCK MINIMIZATION

5.1 The Roles of Man and Computer

Condition that must be observed in formulating the production schedules of press shop may be broadly classified into the following two groups:
1) Conditions which the schedule operators are not allowed to change arbitrarily.— Restricting conditions such as the production schedule of the subsequent welding shop, the maximum capacity values of the press machines, and the maximum available workers.
2) Conditions which the schedule operators are allowed to change.— Variable condi-

tions:
1. Initial assumed input conditions.
2. Precedence relation.

In order to always grasp current stock size in a proper time series relationship between two shops, a huge amount of imformation must be quickly processed, and for this type of information processing, computer are well suited. However, computer must be used, needless to say, as an aid to man, who is capable of understanding and evaluating the actual situations in the shop, in his task of schedule formulation. The initial assumed input conditions of 2)-1 mean data such the compulsory line working hours, safety time allowance within the operating hours, the process cycle time of each parts and the time for die change, which are regarded as the initial computer input data. As these input data will be found to fit well into the system through conversation of man and computer, they are fixed as restricting conditions. This means that when the output corresponding to some input data is evaluated by man, and he finds no contradiction in the production process and positive merits on the cost and the working load of the workers, these input data will be accepted as restricting conditions of 1). Then, the factors of the precedence relation of 2)-2 such as lot size and process sequence will be processed as variables. In the proposed method, man assumes the leading function by establishing the computer input conditions, modifying and adding conditions in the course of processing, and evaluating the output on the basis of his stored knowledge and his evaluating faculty, while computer assumes the function of quantitative processing, logical computation and rapid display of the computation results.

5.2 The Application of The Method of Stock Minimization

The processing method of input and output data related to the fluctuations in time series of stocks and their containers, and the die changes of production line, as examined here will also apply generally to all stock minimization problems between any two different job shops such as between a paint shop and an assembly shop.

Fig.8 shows a flow chart of the method by the man-computer system. In the flow chart, "DISPLAY" means where data are displayed on the CRT, for man to evaluate the appropriateness of the prior input conditions. On the basis of his judgement, he determines whether to
1) Go ahead
2) Go back to setting up variable input conditions
3) Reset the schedule.

The system is also capable of changing the CRT display data for their re-input.

As a simple example, the application of our developed method to the production system of an automobile body is shown in the following. Here, press shop has 2 lines, one of which is composed of 5 press machines and the other is 6. The kind of parts processed in this shop is 200, and that of a welded body which is assembly by these parts is 150. And the average number of parts used per a welded body is 15. These come into production conditions. It is the purpose of this example to calculate the average fluctuations of in process inventory (stock) in three months when two factors of lot size and process sequence change respectively, so that it is possible to examine the effect of two factors on the stock. We set up 2 classes of 6000 and 4000 for lot size, and 5 types of process sequence with regard to each lot size, as shown in Table 3. The groups of common parts is also assumed 3 types of X,Y and Z, and the number of production of X is 8000 per a month and that of Y and Z is 4000 per a month. Fig.9 shows the results obtained by this method, based on the above assumptions. These represent that, although the quantity of lot size strongly effects on the average of stock, stock for R_0-1 or R_0-2 of lot size 6000 is less than that for R_0-10 of lot size 4000. This means that the stock also depends on process sequence. It is a desired process sequence that represses the increase of stock and improves line efficiency by fewer die changes of a line.

6. CONCLUDING REMARKS

In this paper, the contents of our developed methods and their applications have been discussed in relation to examples taken from the problem of line balancing and stock minimization of the assembly line. These methods, which have been developed to make for the defect of the conventional method, combine a subjective judgement with objective scientific data and hence are considered highly significant.

REFERENCES
1. Kawabata,M. "Assembling System and Computer", Sangyo Tosho,Japan,1971.
2. Warfield, J. N. "Toward Interpretation of Complex Structural Models", IEEE SMC Vol. 4, No.5.

Table 1 Allotment of Work between Man and Computer

Man	Computer
To determine the precedence relation among MRWEs through pair comparisons.	To form the system structure based on the precedencd relation (ISM program).
To evaluate the results of line balancing.	To handle combination problems (line balancing program).
To adjust and modify the restricting conditions.	To compute various values, such as parameters and to output the results.

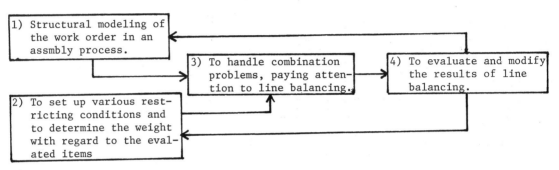

Fig.1 Formation Diagram of the Problems of Line Balancing

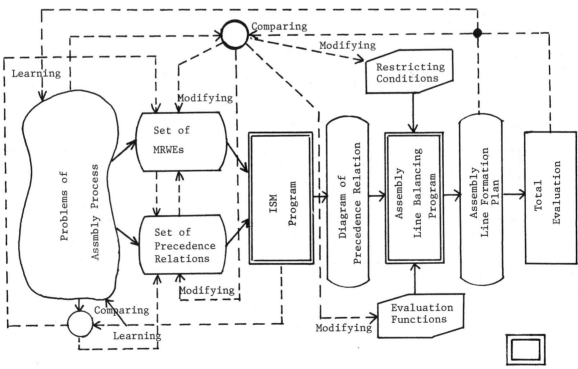

Fig.2 Formation Diagram of Interpretive Structural Modeling of Assembly Process (ISMAP)

Processing by Computer

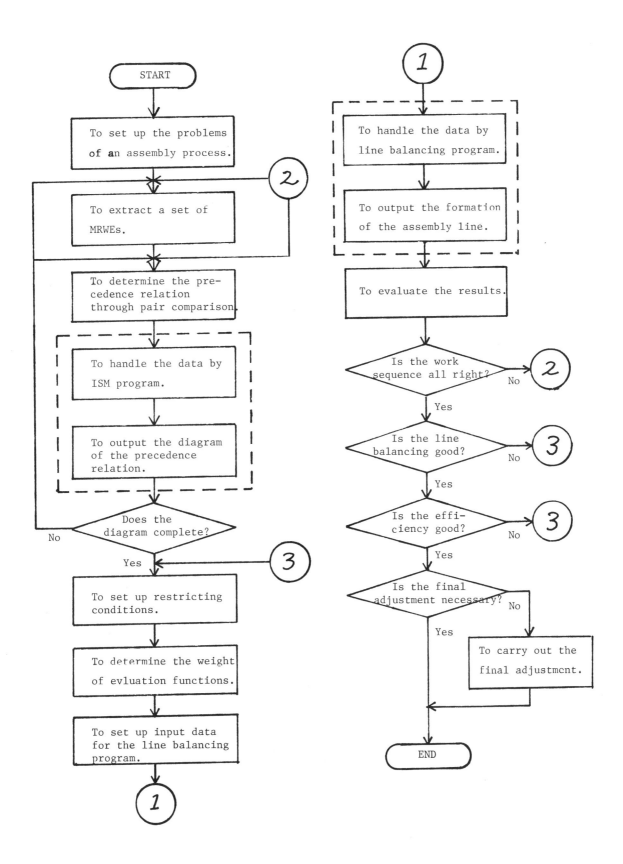

Fig.3 Flow Chart of ISMAP

Table 2 Each Condition of MRWEs

MRWE's Number	1	2	3	4	5	6	7	8	9	10	11	12	13	14	15	16	17	18	19	20	21	22	23	24	25	26	27
th(J_x)	5.	8.	4.	5.	4.	10.	4.	3.	13.	8.	3.	6.	4.	4.	7.	5.	4.	4.	13.	6.	12.	6.	8.	4.	13.	8.	4.
m(J_x)	1	1	1	1	1	1	1	1	1	1	1	1	1	1	1	1	1	1	1	1	1	1	1	1	1	1	1
Pos	R	R	L	L	L	R	R	L	B	B	R	R	L	L	R	R	L	L	L	L	B	R	R	R	L	L	L

(Note) th(J_x) ---- Standard working hours for MRWE J_x (sec)
m(J_x) ----- The number of workers required for MRWE J_x
Pos ------- Working position of MRWE J_x (R - right side L - left side
B - right or left side)

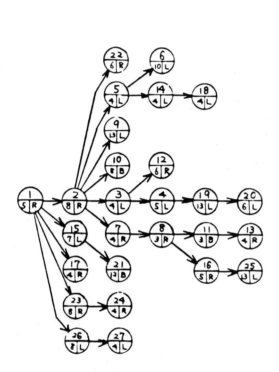

Fig.4 Diagram of Precedence Relation among MRWEs

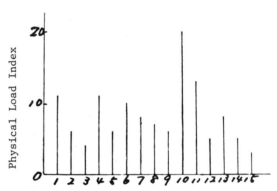

Fig.5 Physical Load of Each Worker (1)
(No attempt is made to balance the physical load of workers.)

Cycle time = 18.0
The upper limit of the rate of operation = 0.8
$\alpha_1 = 1$
$\alpha_2 = 0$

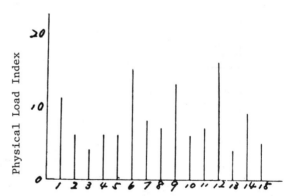

Fig.6 Physical Load of Each Worker (2)
(An attempt is made to balance the physical load of workers.)

Cycle time = 18.0
The upper limit of the rate of operation = 0.8
$\alpha_1 = 1$
$\alpha_2 = 0.2$

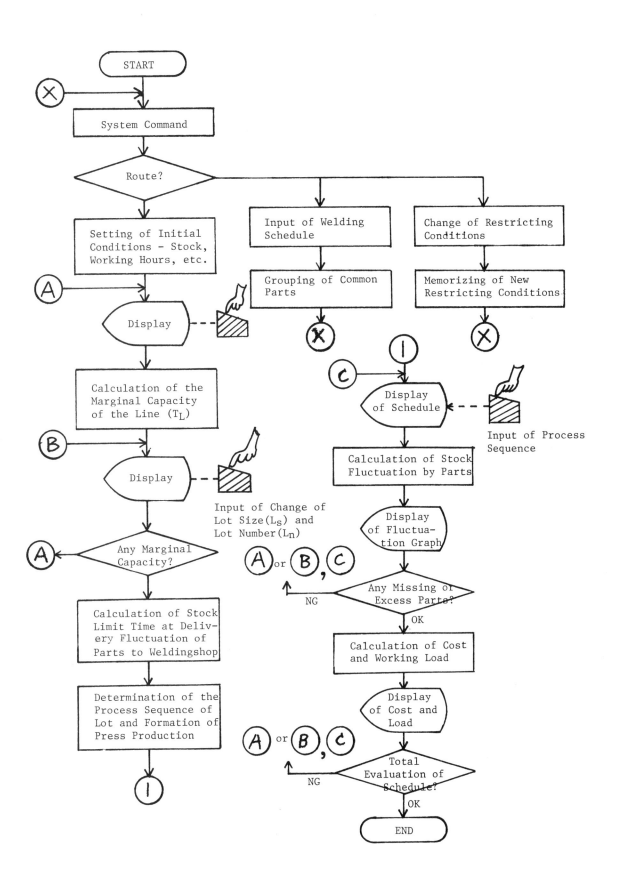

Fig.8 Flow Chart of Method of Stock Minimization

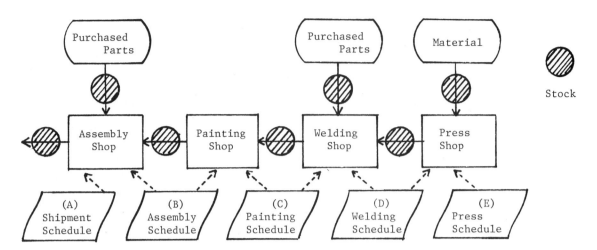

Fig.7 Production Process of Autombile Body

Table 3 Lot Size and Process Sequence

Data No.	Lot Size	Process Sequence
$R_O - 1$		X,Y,X,Z,X,Y,X,Z,----------
$R_O - 2$		X,Z,X,Y,X,Z,X,Y,----------
$R_O - 3$	6000	X,Y,Z,X,Y,X,Z,X,----------
$R_O - 4$		X,X,Y,Z,X,X,Y,Z,----------
$R_O - 5$		X,Y,Z,X,X,Z,Y,X,----------
$R_O - 6$		X,Y,Z,X,-----------
$R_O - 7$		X,Y,X,Z,-----------
$R_O - 8$	4000	X,Z,X,Y,-----------
$R_O - 9$		X,X,Y,Z,-----------
$R_O - 10$		X,Y,Z,X,X,Z,Y,-----------

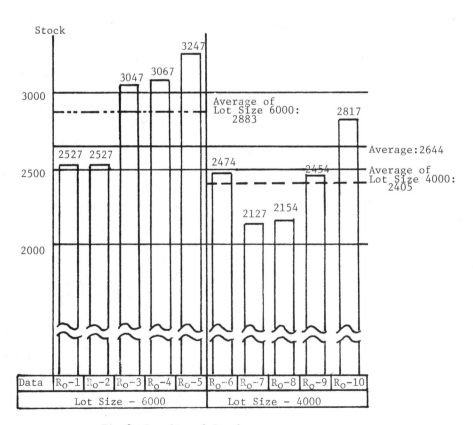

Fig.9 Results of Stock

5. LATE PAPERS

TWO CASES FROM THE NORWEGIAN PROCESS INDUSTRY

A. B. Aune

SINTEF, Division of Automatic Control*, 7034 Trondheim-NTH, Norway

INTRODUCTION

Since the early 60's Norway has shown a rising tendency in initiating activities aiming at humanisation of work. Some of these have been research projects, others the preparation of agreements between the employers and the employees organisations or new legislations.

Most well-known are several socio-technical projects named the Norwegian Industrial Democracy Programme. Keywords in this programme were autonomous groups, codetermination of workers in job design, and quality of working life indicators. An increased use of automation, also including computers, took place in industry as well as in office work. The fear among employees of being run down by new technology led to several actions in the employees organisations. The Iron- and Metalworkers Union in 1970 initiated a research project with the aim of studying computer-based production planning - and control systems and the workers participation in system design.

This project contributed to the preparation of an agreement on "The Use of Computer-based Systems in the Norwegian Industry". A major step was taken towards creating a more human working life in Norway, when the new Work Environment Act was put into force by July 1., 1977.

A growing understanding among automatic control engineers of human needs in automated systems, in 1972 led to a research project named DUPP (Partly Unmanned Production Processes). This project was aimed at: modular and easy-to-use control systems, study of social effects of automation and the use of automation in order to reduce the need of shift work.

After two years of discussions, misunderstandings and some active work one managed to create two new projects - the two projects to be presented in this paper.

Both projects are based on a cooperation between research organisations, process industry and the Royal Norwegian Council for Scientific and Industrial Research (NTNF). The financial support is shared by industry and NTNF.

CASE ONE: PVC-TUBE EXTRUTION PLANT

This is a plant where PVC-tubes, of many dimensions, are manufactured in parallel extrution units.

The total plant is divided into four sections (Fig. 1):

1. The mixing section

PVC granulate and several additives are batch-wise weighed, mixed, heated and cooled and, in form of new granulate, conveyed to a row of feeding silos.

2. The extrution section

From these silos granulate is fed to extruder units. These extruders continuously produce tubes of a certain diameter. After leaving the extruder the tube is water-cooled and automatically cut to the correct length.

3. The muff coupling section

From an intermediate store the tubes are brought to units where one end of the tube automatically is formed as a muff coupling.

4. The storage section

After the muff forming is finished the tubes are counted, bundled and stored before they leave the plant on their way to the customer.

Important quality factors, such as burning spots, tube wall thickness, tube diameter, roundness and strength factors, are checked manually or automatically.

---

* SINTEF The Foundation of Scientific and Industrial Research at the University of Trondheim.

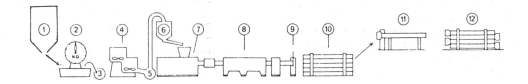

Fig. 1. Production of extruded PVC-tubes.

1. Raw material silos
2. Automatic batch weigher
3. Automatic tube conveyor
4. Mixing units
5. Automatic tube conveyor
6. Feed silos
7. Extruders
8. Water cooling units
9. Cutting units
10. Intermediate tube store
11. Automatic muff coupling units
12. Finished tube storage

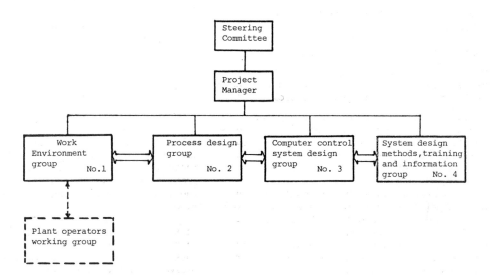

Fig. 2. Project organization with co-operative activity groups.

Project description

A R&D project was started in 1974. Its main goal was to develop a multi-computer control system for supervision and control of the production. This development should be carried out in close cooperation with the plant management, plant operators and their union. Furthermore should be studied the possibility of reduced need of shiftwork by utilization of the new control system.

The total project was organized as shown in fig. 2. The various working groups have participants from research institutions, plant management, plant operators, local union representatives and the computer system vendor.

The goals of this project seek to:

o Find methods and techniques which make utilization of computerbased control system more feasible to Norwegian industry.

o Analyze the influence a high level of automation have on work environment, and find ways to create a more human work environment by means of automation.

o To increase the competitive power of Norwegian computer control system vendors.

The total project work is split into the following activities (Fig. 2):

1. Automation and work environment
2. Automation of the manufacturing process
3. Computer technology as tool in automation
4. System design methods, training and information

The main technical activities are:

o To install new and to improve existing process equipment. This enables the planned utilization of the computer control system.

o To design and install a multi-computer process control system as shown in Figure 3. The final set-up is a result of two years of cooperative design work by the project subgroups (Fig. 2).

The technical solutions, being results of an interdisciplinary design work, are implemented by the technical sub-groups in a traditional way. The following presentations is mainly concerned with the Work Environment group and the practical problems centered around interdisciplinary system design work in practice.

Automation and Work Environment

The group responsible for this activity is composed of:

o A company management representative

o The computer Shop Steward (A new job as a result of the earlier mentioned agreement on "Use of computers in industry") elected among the plant operators.

o Project manager

o A social Scientist

In matters concerning plant operators the group is backed up by an additional group of plant operators. Furthermore, representatives from middle-management, the laboratory and office workers frequently attend meetings of the Work Environment Group.

In the long run a project of this type and dimension will have an effect on all company activities. There is, however, no doubt that the major changes of work will be felt by plant operators.

The work in this group has mainly been concerned with

o A survey of the physical and psychical factors of the work environment before the introduction of new technology and organization.

o An analysis of consequences through continuous evaluation of work environment aspects of technical proposals presented by the other working groups. Analysis is supposed to be done as a parallel activity to the technical development, thus enabling a real codetermination before technical solutions are "frozen".

Work environment is defined as a set of environmental factors. The factors which were found to be of greatest importance in this project are

· Physical environment
· Jobs, job content and the physical environment
· Shiftwork
· Organization
· Co-determination
· Training

All proposals presented during the systems design phase are evaluated with respect to these factors. Up till now, the greatest problems of such work have been related to the organizational changes on the shop-floor.

In this factory there will be a major change in the arrangement of work for the shift-workers (the majority of employees). At present, shiftwork is carried out by 7 operators, each skilled for work in one of the production areas (Fig. 4A). The plant is manually operated via consoles on the shop floor.

The future arrangement of work will need only 4 operators, supervising the total plant mainly from a new control room. The control room is equipped with consoles for automatic computer control, colour TV-monitors and manual back-up stations. Still some instrumentation will be located on site

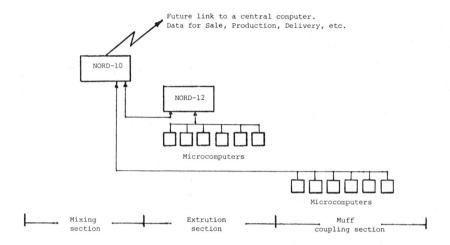

Fig. 3. Hierarchic configuration of the computer control system.

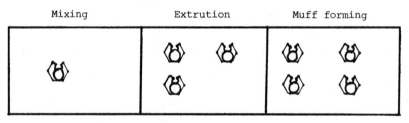

A) Present situation

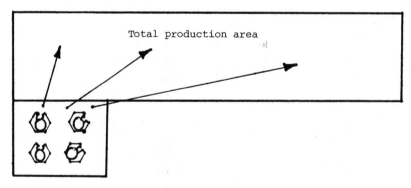

B) Future situation

Fig. 4. Organization of shiftwork groups.

in the production area in order to perform local control and maintenance. The shift working teams will have rotating jobs and work will be planned and organized by themselves.

What have we learned from this project?

The project is now in the middle of the implementation phase. It is, therefore, not possible to present final results.

Nevertheless, based on SINTEF's involvement in the project, by representatives from the Automatic Control and the Industrial Social Science groups some comments may be appropriate.

o Comprehensive automation will have a strong influence on work environment and organization.

o Comprehensive automation in most cases will lead to reduced need of manpower.

o In this project, workers becoming redundant on shiftwork will be occupied with other jobs in the company. Reduction of manpower down to the "natural level" of the future system will take place through "natural reduction" (e.g. retirement, etc.)

o Interdisciplinary systems design is time-consuming and, therefore, somewhat in opposition to traditional industrial-economic thinking. It is easy in theory and problematic in practice. Practical systems design work is strongly influenced by attitudes and human and political views. Real co-determination calls for a change in attitudes, both among employers and employees as well as among systems design experts.

o Co-determination also calls for a "good" method for communication. In this project a combined block diagrammatic-verbal description method was used. Although, not used in full, this method added a new dimension to the design of a transparent system, and thus, to a better understanding of the system.

o Work organization is also easier in theory than in practice. The plant operators feel a great need for training before they go into the new organization.

o When working with job design, work organization and evaluation of work environment criteria, one often find little training among people to "picture" future work situations (model thinking).

o Reduced need for shiftwork has been discussed. After some experience with the reliability of the control system and process equipment this matter will be considered again.

CASE TWO: CEMENT PLANT

Early 1976 SINTEF started a project named "Social consequences of Automation" with grants from The Royal Norwegian Council for Scientific and Industrial Research (NTNF).

The main goals of this project are:

o Through several co-operative projects together with employers, employees and their interest organizations to create a general insight into the social consequences of computer-based control systems in industry.

o The results should help the involved groups to a better understanding of the problem in general and of their own work situation, thus improving the conditions for influencing the development of work and work environment.

o The results are expected to have an important influence on research and education in the field of automation.

A very important part of this project was to establish contact with the Norwegian industry in order to find one or more enterprises interested and willing to start a co-operative research project with the research group at SINTEF.

After negotiating with several enterprises, a project agreement was reached with the largest cement plant in Norway.

Plant description

Figure 1 shows the principle of the cement plant. The total production is divided into four production departments.

1. Quarry and mine, stockpiles

Limestone is broken in a quarry and a mine and conveyed to two large stockpiles where a homogenization take place. The size of limestone is reduced by secondary crushers before material is fed to silos in the raw-meal department

2. Raw-meal department

Various qualities of limestone are blended with additives (quartz and iron oxide) to form the correct chemical composition of the material. Material is first ground in an Aerofall mill. It leaves the mill in two streams. The coarse material stream goes to several ball mills, where it is reground to a fine powder material. This stream meets a second fine material stream from the Aerofall mill. Together they are conveyed to a set of silos, homogenizing silos, where the final blending take place.

3. Kiln department

Raw-meal is then fed into four rotating kilns and burned to clinker (pellets).

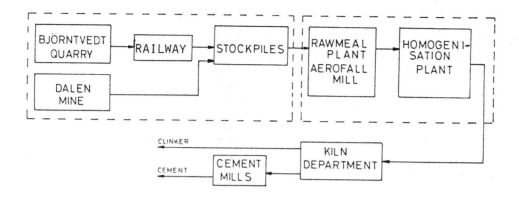

Fig. 1. Block diagramme of the cement plant showing the four plant departments.

4. Cement mill department

Clinker from the kiln department enters a new set of ball mills, and is, with additional gypsum, ground to finished cement of various qualities.

A project description

Before reaching an agreement with the cement company, several demands had to be met:

o The project decided upon, had to be of interest to company management and the plant operators and their union as well.

o A local project group with representatives from the interesting parties should be formed and act as the local "driving force".

o Company representatives should be given time to work in the project as much as possible, but without serious neglection of their daily work.

All these items were agreed to and a project plan was drawn by the two co-operative groups having the following composition:

The company group

o One management representative
o The Raw-meal Department foreman
o A Raw-meal Department shop steward
o The local union secretary

The SINTEF group

A research fellow from each of the groups: Automatic Control, Industrial Social Science and Industrial Engineering.

The project should be carried out within the Raw-meal Department, where several major steps in automation and organization had taken place since 1965.

This is a production area where control and supervision is carried out in three separate control rooms

1. One for the Aerofall mill
2. One for regrind mills
3. One for supervising the operation of the homogenization silos and the feed of raw-meal to the Kiln Department.

Control room No. 1 was erected with the Aerofall mill in 1966. This system led to a higher level of automation. The lod regrind mills and control room No. 2, from before 1965, were modernized. Plans were made for a new Aerofall mill in 1970, with one central control room for the total Raw-meal Department. For market and economic reasons these plans were put aside. Control room No. 3 also was modernized in 1966 with the introduction of the Aerofall mill.

Quality control is completely dependent on a good operation of process units belonging to all three control rooms. Obviously, having three separate control rooms may cause practical operational problems. Up to 1973 a local chemical laboratorium was situated in control room No. 3. Samples of the material streams were manually taken and analyzed in this lab. (An hour to hour quality control in co-operation with the central chemical lab.

In 1972 an automatic material-stream-sampling unit was installed with automatic conveyance of samples to the central lab. In 1973 the quality control was reorganized. All quality control tasks were transferred to the central lab and centred around an automatic X-ray analyzer and an on-line computer.

The total manning of the Raw-meal Department was reduced from 30 in 1965 to 22 in 1976. This happened in spite of an increased production and a transfer from a 4 shift to a 5 shift arrangement.

A two-piece project was formed in order:

1. to survey the development of automation and organization over the last 10 years (1966-76)

2. to utilize collected knowledge and views as a basis for future development of supervision and organization of work in the Raw-meal Department.

Briefly, the key activities of the project are:

1. **Development of automation and organization of the Raw-meal Department in the period 1966-1977.**

   I Surveying
   - changes in technology, process equipment and production methods
   - changes in organization, manning and work conditions
   - changes in work environment

   2. Planning and accomplishment of changes
   - Reasons for changes
   - The planning process. Iniative from whom?
   - Accomplishment. Co-determination, by whom?

   3. Viewpoints on the development, among
   - employees in the Raw-meal Department
   - planners/technical staff
   - company management
   - the labour union

   4. Evaluation of
   - planning and development work
   - forms of co-operation
   - organization
   - social conditions/work environment

   II Future development

   1. Existing plans for future development of supervision and organization of the Raw-meal Department.

   2. Surveying requests and views on future development, among
   - operators of the Raw-meal Department
   - planners/technical staff
   - company management
   - the labour union

   3. Proposals for future development, including evaluation of
   - work conditions
   - work environment
   - organization/task allocation
   - supervision and control tasks
   - control room design, man-machine communications

The first part of the project is now finished. Data have been collected by means of questionnaires and personal interviews. The SINTEF group had to perform part of their work on a shift basis.

Comments to the project

o Introduction of the highly automated Aerofall mill caused more negative attitudes among the oldest workers than previously assumed by co-workers and management. However, the speed of changes were never felt as a serious matter among most of the present operators. (Most of them have had various jobs in this Department in the period considered).

o Very little co-operative work was brought into systems design. It has been a very traditional (in an international meaning) way of design and implementation.

o A total supervision of this Department by one operator in each control room is not ideal, especially since quality control is performed on a lab to control room No. 1 basis. Socially seen, the reduced possibility of personal contact between operators was not felt as a serious problem.

o The transparency of process operation and quality control, is little to the operators. It means that quality figures as results of the operators' actions are not fed back to the operators.

o Among the operational people of the Raw-meal Department the present manning is considered as too low. This is a very stressing situation. When trying to survey view-points on future development, one is frequently met be the fear of further reduction of manning. Job security is also an important factor in this discussion.

o The labour union has no clear views of how they should act with respect to a future development based on more automation. Matched to an understanding of further reduction of manning, is the acceptance of this being possible only by an increased use of automation. (The competition problems on a world-wide market). This can only be accepted with a strong co-determination in systems design and pace of development.

o So far, the project work has been very promising. It has been much more time-consuming than expected. Even with un-restricted financial limits, an industrial project can only be paced in a way which matches the normal life of production. Finally, motivation of and active co-determination by shift working people is in general a practical problem.

DISTRIBUTED CELLULAR MANUFACTURING SYSTEM

A. B. Aune

SINTEF, Division of Automatic Control*, N-7034 Trondheim-NTH, Norway

INTRODUCTION

Manufacturing technology has brought about an increased standard of living, but also human and environmental problems. In more recent years, manufacturing technology has often become a double-edged sword. The introduction of numerical control (NC), which is one of the most important events in manufacturing history, increased productivity tremendously, but often led to the skilled operators' job becoming dull and boring. Much of the machine tool operators' work was transferred to specialist groups (e.g. production planning, machine settings, tool selection, maintenance, etc.).

The development of numerical control systems for machine tools is developing rapidly (Fig.1). In the past, human and environmental disadvantages were accepted as a necessary price of economic benefits. Influenced by what one may call an international human/environmental revolution, manufacturing engineers now have to be more conscious of human and environmental problems created by new technology.

This is strongly emphasized in the new Norwegian Work Environment Act, where objectives are:

1. to secure a working environment which affords the workers full security against harmful physical and mental influences and which has safety, occupational hygiene and welfare standards that conform with the technological and social structure of the community at any time,

2. to secure sound contract conditions and meaningful occupation, for the individual worker.

3. to provide a basis whereby the enterprises themselves can solve their working environment problems by collaborating with labour organisations under the supervision and guidance of the public authorities.

This paper presents the Norwegian project on "Computer Managed Parts Manufacturing" (CMPM) with the work emphasized on cellular manufacturing, utilizing the principles of group technology. The project is sponsored by the Royal Norwegian Council for Scientific and Industrial Research (NTNF) and is carried out at the Production Engineering Laboratory, SINTEF.

The purpose of the project is to develop methods for increasing the level of automation of small batch production in the mechanical industry. This can be done by means of grouping a number of NC-machine tools together with equipment for automatic handling. The control is given to an executive system where minicomputers play an important part. The expected results of the project is a modular system consept, both on the handling side and the system side, which enables the user to make simple or complex solutions according to his own situation.

However, in contrast to similar projects in other countries, one is not aiming at the "Unmanned Manufacturing" concept. Instead, the philosophy behind the project is built on elements from cellular technology, self-controlled groups, and decentralized production. A very important matter is the analysis of the operator's situation in a more automated system concept, and an adaptation of the technology to the role he wants in the system. As the project is seen to have influence on the job situation in the mechanical industry of tomorrow, it is watched carefully by, and partly carried out in cooperation with the Trade Unions Council of Norway.

THE CONCEPT OF CELLULAR MANUFACTURING

The types of production systems most commonly used within workshop industry are:

- Production lines
- Functional groups

In a production line the machines' tools are grouped along a line in an order corresponding to the operations performed on the product.

---

* SINTEF The Foundation of Scientific and Industrial Research at The University of Trondheim, Norway.

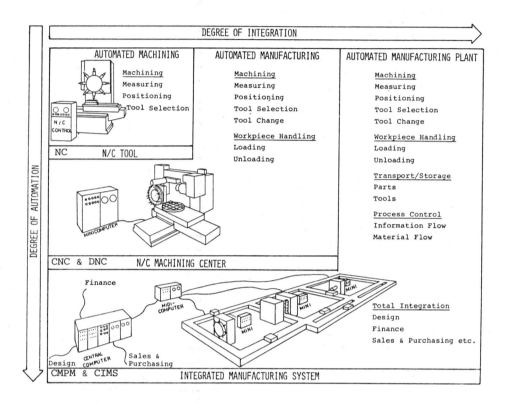

Fig. 1. The development of numerical control systems for machine tools.

```
NC   = Numerical Control
CNC  = Computer Numerical Control
DNC  = Direct Numerical Control
CMPM = Computer Managed Parts Manufacturing System
CIMS = Computer Integrated Manufacturing System
```

Functional groups mean that machine tools capable of performing the same type of operations (i.e. turning, milling, grinding, etc.) are grouped together.

Lately, a new concept, cellular manufacturing has been used more often. A set of machine tools necessary for the production of a family of similar parts, are grouped together. This group is called a manufacturing cell. All operations on a given part are performed in a straight sequence with no unnecessary delay between each operation. In this way lead time is strongly reduced compared with the "functional group" systems.

A number of cells can be situated within the same factory.

One advantage of the cellular concept is that production facilities can be divided into cells and spread geographically around the provinces where the potential of workforces are living. Advantages may be a stable and motivated workforce, even with some positive effects associated with the concept of "small is beautiful". Disadvantages might include absenteeism and additional transportation of products.

In order to handle dataflow between the cell and the parent company, a well developed telecommunication system is needed.

A typical cell would have five functions:

- Planning
- Preparatory functions
- Manufacturing
- Supplementary functions (e.g. quality control, etc.)
- Stocking function

Manufacturing is performed in the cell-core, a number of numerically controlled machine tools. A cell-core also would consist of an automatic workpiece handling device, an industrial robot. The robot moves the workpieces both between machines and between machines and the corestock. The corestock has a capacity to handle the number of workpieces machined during two shifts. Thus, the cell-core is working around the clock unmanned, while the cell as such is manned only during daytime. Each cell should consist of about 10 operators with some additional help from outside the cell.

CELL REALIZATION AT SINTEF

As part of the research project a pilot version of a cell-core is built up in the Production Engineering Laboratory of SINTEF (Fig. 2).

The main activities around this cell are:

- to establish the pilot version of a cell-core in the laboratory

- to design and develop planning and manufacturing systems including also man-machine communications

- to operate the cell with skilled operators in order to evaluate the users' attitudes and acceptance.

Pilot production will take place in 1978 in co-operation with a Norwegian shippropulsion systems manufacturer. The parts to be manufactured are selected mainly from diesel engins. Parallel with the pilot production, one also hope to start planning of the implementation of two real cells which will be placed out in the province. If the results are positive conditions may become suitable for a large scale industrialization of the Norwegian provinces.

On the hardware side, the production cell contains 4 different CNC (Computer Numerical Control) machine tools together with a CNC industrial robot (Fig. 3) with a load capacity of 135 kg. In addition, there will be a direct access stock of parts for the robot and a transportation arrangement. Control data is distributed from a central programming center based on a DNC minicomputer.

A very important point in the hardware concept is the development of a product-carrier, the pallet. The different products are manually clamped on these, the pallet serves in the production process as a standard interface towards the hardware side of the system, that is towards the transportation and stock devices, the robot gripper and the clamping in the machine tools. The pallet must have very low weight because of the load capacity of the robot, at the same time as it must be rigid enough to take up the cutting forces. A general purpose pallet for robot handling is, as far as one can see, not realized before.

Closing remarks

o Cellular Manufacturing Systems seem to be a sound foundation for further development in the workshop industry which corresponds to modern principles for the quality of working life. There is, however, no doubt that this development show a possible trend towards "Unmanned Manufacturing Systems". Of major importance is, therefore, the users' participation in systems design work. Only in this way, can good work places in a sound work environment be achieved.

o The Norwegian Iron- and Metalworkers Union have evaluated problems created by numerical control of machine tools in the Norwegian workshop industry. From the point of view that technical development in this industry cannot, or should not, be stopped, they have presented several actions to be taken in order to create human work places (new ways of training and organization of work, etc.).

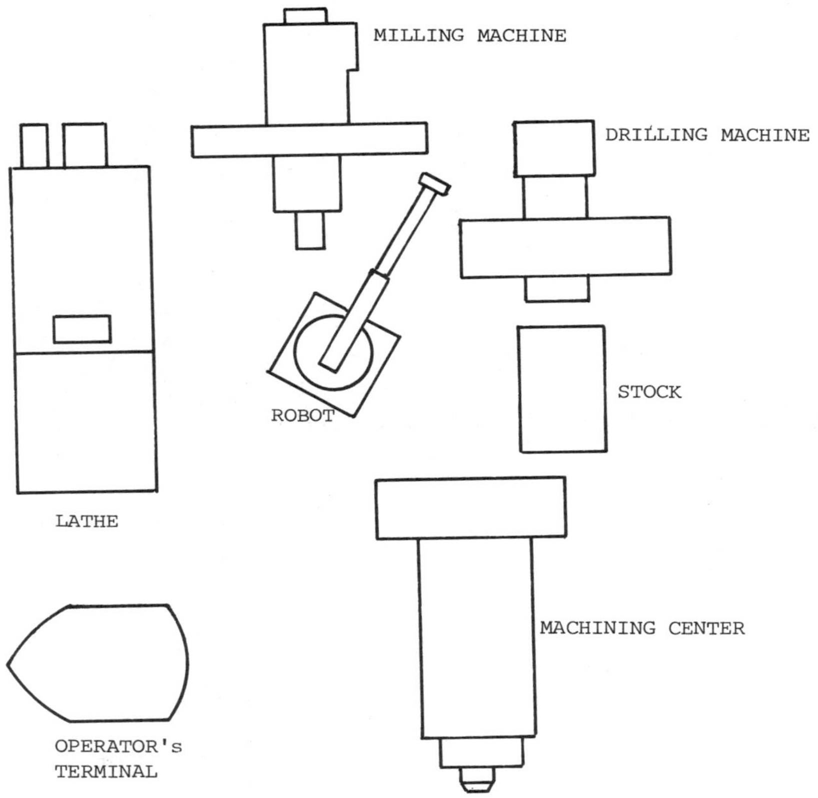

Fig. 3. Full scale cell core and industrial robot servicing four numerically controlled machine tools.

Figure 2. Configuration of the automatic cell built at the Production engineering Laboratory of SINTEF.

AUTOMATION AND HUMANIZATION AT THE DUTCH RAILWAYS; SOME EXAMPLES

D. P. Rookmaaker

Secretary of the Ergonomic Steering Committee N.S.

INTRODUCTION

In this discourse I shall try to describe how the automation projects specifically and the automation strategy in general, are approached by the Dutch Railways (N.S.).
I shall try to give a more or less, detailed picture bearing facets which are connected with the humanization of work or the assignment. The outlined picture tries to be an honest analysis of the reality, where possible supplemented with future expectations as seen through the eyes of the author. Analyses and the suggestions made, are therefore, of a personal vision.
However, I hope to give a realistic, detailed picture of a movement which is already in "full swing" in the organisation of the N.S.
To illustrate this is the fact that at the time of writing, a concept-plan has been conceived which gives a rough, global but complete and mutually consistent futuristic picture concerning the automation 1980-1985. The goal of this so called 'outline' is a social, economic and technological acceptable starting point for a thorough dialog between management, personel and others over the framework of automation projects; their mutual connections and their situation in time. As such, this 'outline' should be able to function as a concrete reference for the automation policy of the N.S.

THE CONCERN 'N.S.'

What is this for a concern, the N.S.?
N.S. is the national rail-transport concern in the Netherlands with an annual turnover, qua receipts, of 1818 million guilders (in 1976). The concern 'produced' in that year 8218 million passenger kilometers and 2.7 million ton kilometers. This was achieved on a total net-length of 2825 km. with 329 passenger stations and 155 loading and unloading yards. This work was done by nearly 27000 personel, who averaged in 1976, 1674 nett work-hours each.
Although N.S. was founded judicially in 1917, the first train appeared in 18.39 between Amsterdam and Haarlem, a distance of 19 km. For the 4 trains to Haarlem and the 4 back to Amsterdam, the time-table could easily be printed on the back of a packet of cigars. The time-table now consist of a book made up of about 500 pages with about three-quarters of a million letters, numbers and visual aids with a yearly circulation of nearly 450.000.
Not withstanding this appreciable growth in nearly 140 years, the N.S. has not been one of the most foreward branches of industry in our country. The nature of the work as well as the method of work, the conception of and the introduction of railway-technology were probably the cause of this.
Just like many other concerns since the second world war, the N.S. has found itself in very strong technical rapids. By the N.S. this was stimulated through:
. Declining profits leading speedily to losses which, in the mean time, have been characterised as structural.
. Rapid technological developments and their proven reliability (zero defects, mean time between failure, fail-safe systems).
. The demand to function more effectively in competition with road-, air- and watertransport.

From this appears a picture of a concern that, due partly to the nature of the work, was, and was able to be, labour intensive. Within a short time, a distinct change can be seen in this portrait expressed in the introduction of (semi) automated systems that are entering into the different branches of the concern.

THE AIM OF THIS NARRATIVE

It is the aim of this narrative to discuss and thereafter to analyse generally, which criteria or better systems of criterium, those examples of automation that are being planned or have been realised.

This analysis will try to show that the criterium 'humanization of work' is often incidental or plays no part at all. In this discourse I shall show subsequently that that, for example, the ergonomy-although frequently not included in a system- ergonomic fundamental approaches in the begin phase of a project are involved - often referred to in the realisation phase. This whilst the more pure behavioral sciences come into contact with automation projects or facets of this, when they are in the evaluation phase and this more or less incidently.
This will be a plea for the creation of a structure, wherein the different 'sciences' can analyse, together, the different aspects of work processes and systems which cohere within the concern. Which includes the aspect of humanization of the assignment. This would have to happen before the decision is made about the question of whether to automate or not, perhaps partially and then over the how and why of similar projects.
This vision is supported through experience, together with results of other analyses, which show that the degree of man-job relation strongly determines the functioning of automated systems.
This aim of a 'multi-disciplinaire' approach has in my opinion general value and confines itself not only to the N.S.

AUTOMATION INSIDE N.S.

With the recent changes in the Statute of the definitions of the aim of the N.S.-alongside preparatory contributions to government policy about transport, traffic and spatial-planning- the following has been added.
The limited Company shall whilst realising her aim, direct her course of action towards the purposeful advance towards self-development, the well being and prosperity of the members of the working community.
In the project (1) 1980-1985) the influencial trends concerning personel and policy, which are directed to a possible structural solution are discussed, for example:
. the elimination of dirty, heavy, dislikeable and dangerous work.
. automatisation of socially unacceptable work.
As has been stated earlier, the recent ouline plan 'Automation' outlines a vision wherein a social, economic and technological acceptable starting point, mutually agreed, must be aimed for. Against that, is the automation policy as it in fact is based at the moment (2):
. Bring the automation facilities up to a level where an optimum of efficiency in business operation can be continually delivered.
. The speeding up of efficiency in the information process itself.
. The following of strict procedures and discreet judgement of the productivity of the new practice.

As a summary of the above, an inconsistent picture is presented in relation to the automation policy by the N.S. On the one hand is the efficiency more or less the only standpoint operated; while on the other hand a plea is being made for the well being of the worker for example the social unacceptability of certain types of work is also a criterium in operation.

EXAMPLE I

In order to illustrate the above, in this chapter I shall discuss examples of automation projects; one in the planning and one in the realisation phase.
These examples, which are to be outlined in their original situations but also in their development, have been chosen reasonably at random. These examples aim to show the most relevant differences in connection to job-situation as it was, before and after, modernisation, as well as keeping the argument in favour of automation. The first example refers to the production processes, processes of planning and publishing of the trainservice.
The planning of a trainservice is centralized but also de-centralized comprising the total production packet available, taking into account the total availeble ingredients qua personnel, material and infrastructure. In the publication process, the planning details are given to personnel and the general public in the form of a time table.
With respect to the planning of a trainservice one can say that up until today this workintensive process is in principle a 'pencil-paper and ruler' occurance, where by a new time-table is planned half yearly, derived from the original but supplemented with the various changes. This planning process is a manually optimalised process wherein all factors connected with the running of the trains are taken into account. The combined actions of the diverse authorities who, partly central and partly de-central, deliver the facts about the personnel (traindrivers and conductors etc.) about trains and type of locomotive, over the situation of line and points etc. etc. leads via type of iterative approach eventually to the desired plannings product.

After the start of the time-table involved one must repeatedly make adjustments and amendments whereby the above explained combined actions continue to play a roll.
It is obvious that this current planning information must be made known to the different authorities. This happens in different forms, tabelarised (for example time table book) or graphically (via, for example the so called track plan).
By the publication of the planning information the control is of the utmost importance. In the first instance pointed towards the part of planning process and thereafter towards publications, there has recently been decided to automate the most important parts of this process. The goal of this automation has been decided by the Management Team for Automation (MTA), a heterogeneous composite group of N.S., described as follows:
.The improvement of productivity per man
.The improvement of the quality of the product
.The adjustment of the mental taxation to the capacity of the personnel hereby taxed, including the rejection of disagreeable work.
.The elimination of manual work, wherefore no personnel will be available in future.
The automation of the planning will include the automation of the administrative and creative part of all plannings activities for all branches. The automation of publication shall partly be made up of computorised plotting of 'time-line' diagrams, tables, printed schedules for drivers and conductors and out of the 'output' which is useable for the computer steered photo-print apparatus of the printers.
Tabel I shows some of the most important characteristics of the present and planned set up together and also the considerations which lead to the decision to automate.

**
The considerations mostly applied are:
-Work saving
-elimination of bottlenecks through shorter completion times
-better quality
-elimination of manual work
-shedding of disagreeable work
-greater flexibility in time and place

ERGONOMIC AND MOTIVATIONAL ASPECTS

How do the aspects which are directed toward the humanization of the job play a part?
With this question we arrive in the territory of the social sciences and of ergonomy; for both of these exist within the N.S.-organization groups which are explicity concerned with this department. The social-psychological approach is mainly organized inside a divisional sector, containing mostly sociologists and psychologists. I shall enlarge on the ergonomy. The organization of the ergonomy, which has already been firmly anchored in the concern for 12 years now, reflects the view that ergonomy requires a multi-disciplinary approach, that the ergonomy in a concern stands or falls with the degree of acceptance of the instructions and finally, that ergonomy is an aspect inside the concern's policy.
Policy questions referring to ergonomic dealings are dealt with in the Ergonomix Steering Group, under the chairmanship of a director and including representatives of the most important commissioners inside the N.S. The execution of the research applications are dealt with by the Ergonomic Team (E.T.), a multi-diciplinary team with technicians, psychologists and work physiologists. The activities of the E.T. occur nearly always only upon request.
The strategy of the ergonomy inside N.S. is strongly directed toward the so called system ergonomical approach.

Present system	Future set up
-Experience and insight of much importance.	-Experience and insight stay important.
-Very complicated pattern of consultations between the different branches.	-Planning centralised and decentralised.
-A long completion time.	-A short completion time.
-Solving problems of a complex nature with only simple help.	-Working in dialogue via videos with the computer system.
-Changes passably difficult.	-Changes passably simple.
-Very time-wasting drawing work.	-Drawing work greatly reduced
-Chances of mistake passable.	-Changes of mistakes minimal through working from one basic proof.
-Work qua supply irregular.	

**

That is that both man and machine are seen as parts of a total system.
The goal of the ergonomical activities is based on the principle that both component parts can be related to each other through acceptance in the direct work surrounding, the job, the machines and the apparatus. This means that after a close analysis of the subject, man-machine systems an allocation, an allotment of job-parts would have to follow according to the possibilities and impossibilities of the man and the machine. By this choice the ethical and social facets would have to be taken into account. In general by allocation problems it shall be accepted that the man is more flexible but less predictable in his actions than the machine.
It is clear that more and more allocation problems have a bearing on the division of man and computer which, according to Jordan (4) are not mutually compareable but are supplementary to one another and complimentary to one another.
Does an increasing automation lead to a optimal division of assignments?
It would be necesarry to have at one's disposal criteria where by this could be judged. Predominantly the following questions play a part.
.Who does it better, man or machine?
.Waht costs more, man or machine?
.Are functions worthy at man created?

On the worthiness of the functions Jordan (4) agrees by the emphasis that he gives to the motivation which is decisive in the functioning of the operator in an automated system.

> You can lead a horse to water but cannot make him drink. In this respect a man is very similar to a horse. Unless the human operator is motivated, he will not function as a complement to machines, and the motivation to function as a complement must be embedded within the task itself. Unless a task represents a challenge to the human operator, he will not use his flexibility or his judgement, he will not learn nor will he assume responsibility, nor will he serve efficiently as a manual backup.
> By designing man-machine systems for man to do least, we also eliminate all challenge from the job. We must clarify to ourselves what it is that makes a job a challenge to man and build in those challenges in every task, activity and responsibility we assign to the human operator. Otherwise man will not complement the machines but will begin to function like a machine. And here too men differ significantly from machines. When a man is forced to function like a machine he realized that he is being used inefficiently and he experiences it as being used stupidly. Men cannot tolerate such stupidity. Overtly or coverly men resist and rebel against it.

Swain (5) goes even further in his theory over the possibilities of the production processes to minimise human errors and mistakes. Starting from the principle that variability is a characteristic which is inherent to every human action, a mistake occurs when the variability goes beyond borders of acceptability. Swain leaves thereby intentionally the line of thought about the lack of interest and demotivation as the main cause of mistakes by men.
He argues that perhaps more of a safeguard to optimal efficiency and profit would be to accept the man as he is instead of trying to make him what the management would like him to be, and try to influence him in that direction. Swain empases to create jobs which the man can and will do.
Going back to our example of the planning process we can establish that there are several valid considerations for the automation approach. It will be necessary to see in how far the new process set up system-ergonomical qua job allocation justifiably is designed. In order to safeguard this there must be continually argued for an early involvement in the set up in order to influence the practise as much as possible on the weight of the diverse arguements that arise Improvement in the measurable efficiency qua production and quality will often be contradictory to the immeasureable goal to eliminate monotone, disagreeable work. It would appear to be of utmost importance to make obvious that the involvement reaches further than the choice of the workplace or the frame of the total picture. More important is actually by the set up and realisation of this plan to contribute that the human assignment forms 'a challenge' for the operator. A thorough analysis of the planning assignments in the form of observation as well as conversation with the involved elements is necesarry.

EXAMPLE II

After example I, wherein it is made obvious in which method we are striving toward the introduction of system automation aspects by an extensive automation project in the planning phase, here is another example from the production sphere. It concerns here the mastering and automation of the train service.

Mastering the trainservice is a controlled process system. This happens through the train dispatcher. A train dispatcher is situated at every station where points an signs have to be operated. The regulatory posibilities of the train service operator is limited because the effects on the regional and rural character of the train service quickly has consequences for the other stations.
On these occasions these instances must be informed. The mastering of the train service has had a lenghty development. One had in 1938 when the first trainservice started between Amsterdam-Haarlem only very simple sign apparatus. Along the track one could find 21 'watchers' each with a white, green and red flag or lantern whereby on the arrival of a train, they waved the appropriate flag or lantern (6). 130 Years later was decided to create a system for very big emplacements whereby the main routes and side-lines according to time-table can be automatically established, even within a certain margin, when technical disturbances occur. Technical disturbance which occur outside this time margin must be regulated by the train service operator via a terminal entry.
The reasons for automation of the mastering of the train service were as follows:
-Better quality of correction actions.
-Improvement of circulation of information in the process.
-Cost-serving.
-Elimination of disagreeable work.
-Lightning of mental taxation.

The exclusion of mutual assesment of these arguments together with by passing of the ergonomic starting point by the man-machine allocation in the expected set up has given rise to a certain number of problems in the realisation of the automation plan. It was quickly noticed by spot checks that the communication and information between man and machine were causing a bottle-neck. This is true in situations where the process, in main, runs automatically but where the train service operator has to intervene in cases of serious disturbance.
One thing and another has shown that in this project, orginally, a clear analysis in the field of train service operation with the diverse branches thereby joined did not occur.
But also apart from this the job allocation, based on facts such as cost, contents of the job, motivation and technique- were hardly considered.

COMMENTS AND CONCLUSIONS

What can we learn from both of these N.S. examples: the promising approach of the automation of the planning and the confirmation of the wrong approach to the automation of the governing of the train service?
In my opinion is the conclusion justified that a systematical integral approach of the human aspect by automation projects is of little consequence.
Facets of this so as the ergonomy are sometimes partly taken into account. Changes in this situation, are as far as I can see for the better -shall possibly in the future, occur because of the experiences that more regularly the man-machine (relationship) is one of the bottle-necks in the optimal functioning of the automated systems. It is, in my opinion, necessary and possible from the above given confirmations and experiences to reach the conclusion.
a. The question of whether certain processes are considered for (semi) automation are' nearly always decided on by technical and micro-economical arguments and seldom or never by social arguments.
b. The method whereby in the main a process (including joining subsidiaries) is going to be automated, are just as seldom decided by the social implication.
c. Too often the ergonomy is introduced into a project in the realisation phase when the mainlines of another must be accepted as concrete.
The pure social sciences (sociology, psychology) are only introduced after the realisation phase for evaluation or for making the necessary corrections if they are introduced at all that is!
d. Other investigations and publications show that the man-job relation is especially by automated processes one of the most vulnerable and problematic links (qua mistakes) Above all it is known that the work-intrinsic factors (satisfactions, motivation) are directly related. See the Swain theory.

From the above conclusions it would appear necessary more integral and clearer, parallel to the ideas of technique and efficiency, to include the idea of humanization of a job. In connection with this it must be said that this multi-disciplinary approach must surely not occur by the realization of an automated project: in fact the phase where the main lines are thought out and set up is incorrect.

In my opinion one must strive towards a multi-disciplinary screening-group who should be able to consider all coming processes multi-disciplinary screened and then decide on an eventual partial or total automation.

This approach guarantees, in my opinion, a responsible policy in questions of automation without certain facets being over or under estimated something which often occurs in a later phase. It is probably unnecessary to say that article has, in my opinion, general validity and is not only valid in the N.S. situation which I have used as an illustration.

REFERENCES

(1) Meerjarenplan, 1978-1982; edition N.S. februari 1976.
(2) Bedrijfsplan 1978; edition N.S. februari 1977.
(3) Landeweerd J.A.; Mens en veiligheid in geautomatiseerde systemen De Veiligheid -53 (1977).
(4) Jordan N; Allocation of functions between man and machines in automated systems; Journal of applied psychology vol.47 (1963).
(5) Swain, A. Semmar: Pratical methods for reducing human errors in production (1974).
(6) 100 Jaar spoorwegen in Nederland; edition N.S. (1939).
(7) Singleton, W.T.; Theoretical approaches to human error; Ergonomical Research Society (1973).

MECHANIZATION/AUTOMATION AND THE DEVELOPMENT OF THE LEVEL OF EDUCATION AND QUALIFICATION OF GDR EMPLOYEES

H. Maier

Akademie der Wissenschaften der DDR,
Zentralinstitut für Wirtschaftswissenschaften, Berlin, G.D.R.

In the socialist economy of the GDR the quota of graduates of technical schools and universities increased from 6,7% of employees in 1962 to 14,1% in 1975. During the same period the quota of skilled workers and foremen rose from 33,7% to 53% and the quota of semi-skilled and unskilled workers declined from 59,6% to 32,9% (slide 1 and 2).

It already can be foreseen that up to 1990 the quota of graduates of technical schools and universities will increase to about 20%, the quota of skilled workers to about 60%. The quota of semi-skilled and unskilled workers will decrease to about 20% *

In 1975 10% of the young people who entered into working life were university graduates, 15% came from technical schools, 60% were skilled workers and only about 15% had no kind of graduation or vocational training.

Alone in the periode between 1971 and 1975 more than 220,000 graduates of technical schools and universities started working in the national economy of the GDR. By this the stock of graduates of technical schools and universities in the GDR economy increased by more than 30%.

In the same period more than 920,000 boys and girls finished their job training which is based on the ten-form general polytechnical school. With that the stock of skilled workers of 1970 was enlarged by 35%. At the same time large parts of employees who already were active in the working sphere participated in adult education. From 1971 to 1975 370,000 employees got a vocational training by adult education. 65% of skilled workers at the age of 50 to 60 got their vocational training already under socialist conditions.

The increasing importance of qualified labour in the reproduction process can be seen in the rising volume of educational funds in GDR economy. The educational funds are the expenses of socialist society for education and qualification materialized in the qualification level of the employees. The educational funds in the GDR increased from 66,5 billion marks in 1962 to an amount of 150,8 billion marks in 1975. That is about one quarter of the funds of fixed assets of today's GDR economy (slides 3,4,5).

In the period between 1962 and 1965 the growth rate of the educational funds was essentially higher than that of the funds of fixed assets. During this time the educational funds increased to an amount of 227%, compared with an increase of 165% of the funds of fixed assets. The research funds in the material sphere - that are the research expenses materialized in the scientific-technical level of production - increased to an amount of 333,5%. Educational funds, production funds and research funds are the technological funds of a society, which gain more and more importance for the scientific-technical revolution. In 1972 the technological funds of the GDR consisted of 73% production funds, 21% educational funds and 6% research funds** (slide 5a).

The high efficiency of social expenditure for education and qualification can also be seen in the close connection between the increased qualification level and a growing contribution of the innovator's movement to the

---

* Annotation: Klaus Korn/Harry Maier (Editors), Oekonomie und Bildung im Sozialismus (Economy and education in socialism), Berlin 1977, S 184.

** Annotation: H.-D. Haustein: Messung der volkswirtschaftlichen Intensivierung (Measuring intensification of national economy), Berlin 1976, S. 49.

efficiency of the national economy. The benefit of the innovator's movement (more than 50% of prime cost reduction in the national-owned economy of the GDR result from it) per unit of educational funds was 2,5 times higher in 1971 - 1975 than during the period of 1960 - 1965. Alone in 1976 1,6 million employees participated in the innovator's movement with a benefit of 3,6 billion marks.

The increasing importance of qualified labour in the reproduction process is an essential productive and social power. This power must be utilized in a better way than up to now, both for the development of socialist production and for the further formation of socialist way of life. Thereby the demands of production and way of life become more and more interlaced.

The improvement of the structure of the national economy, the higher demands upon the material-technological basis in the further development of the national economy require "to bring to bear totally the advantages of our country, especially the power of his experienced and qualified working people"*
That means to realize a structure policy that utilizes the high qualification level of working people in order to decrease the raw material and energy intensity of production by a higher rate of intelligence intensity. As important problem thereby is the employment of working people according to their qualification. Only by an effective use of qualified labour the productive potency, created by educating and qualifying people, can be transformed into real productive forces development. Only in the labour process the socialist personality can totally be developed. Therefore the aimed employment of qualified people in the national income producing spheres, that means in the material production, has great importance. In this sphere, especially in the industry, 45,2% of the increase of university graduates have been employed in 1971 - 1975. So in this period the stock of university graduates in the industry rose to an amount of 180%. In the whole sphere of material production the increase of this stock was 175% In the same time the stock of graduates of technical schools grew up to an amount of 136,8% in industry and to 138,5% in the whole sphere of material production.

By that industry could nearly double its share in the total of university graduates in the period of 1962 to 1975. In 1962 every tenth university graduate of the GDR economy worked in industry, today it is every fifth. Building industry, transport and posts and telegraphs could even more than double their share in the total of university graduates. A similar development can be seen in the field of employment of technical school graduates (slides 6, 7).

The intensified employment of highly qualified labour in the sphere of material production made it possible to supply such fields as technology, construction, investment projects preparation, research and development, transfer and application as well as steering and regulating of highly productive equipment with highly qualified manpower.

The scientific-technical progress is not only the decisive process for a gradual decrease of unskilled labour but in the same time the key for the further development of the material-technological basis. Though the scientific-technical progress in the socialist society raises the importance of qualified labour in the reproduction process its effects on the qualification level are not without contradiction.

So for instance we find the highest share of semi-skilled and unskilled workers at manual, fully mechanized and partly automated working places. The highest share of skilled workers can be found at partly mechanized and fully automated working places (slides 8 and 9).

It can be seen that even if the requirements to manpower differ widely depending on the level of technical development the basic tendency is the rising importance of qualified labour in connection with the increasing degree of mechanization and automation.

Considering these facts we never must ignore that at present only about 12% of the employees in the GDR work with automated equipments and the share of production workers in industry who do not work with machines and equipments has an amount of about 42%. Half of them has to do hard physical work. Automation brings along higher demands to qualification, but at the same time

---

* Annotation: E. Honecker: Aus dem Bericht des Politbüros an die 5. Tagung des ZK der SED (Report of the Politbureau to the 5. Session of the SED Central Committee). "Neues Deutschland", 18.3.1977, S. 4.

releases manpower that then partly has to work in production processes with lower qualification demands. An essential task is to reduce hard physical work. In the current five-year-plan period from 1976 to 1980 the number of such working places is to be reduced by 25 to 30%. There is a close connection between increase of qualification requirements and decrease of hard physical work. The high share of production workers who do not work with machines and equipments shows the great importance of mechanization in the present stage of development. Mechanization does not only essentially contribute to abolish hard physical and unskilled labour but creates decisive preconditions for automation. The relatively high share of unskilled workers at automated equipments results from the fact that at present automation does not completely cover the whole production cycle. At present mainly direct machining tasks are automated so that feeding and emptying of automatic machines as well as the connection to other sections of the production cycle has to be done manually.

In increasing the importance of qualified labour attention has to be paid not only to mechanization and automation of the main production processes but increasingly to auxiliary and subsidiary processes. Especially here mechanization and automation make it possible not only to release manpower in a large scale but also to increase the importance of qualified labour essentially. This is in particular true for the mechanization and automation of processes of transport and turnover.

The aimed and systematic abolition of unskilled labour is an essential requirement of the further development of the material-technological basis in the GDR. This is the only way how the creative potential formed in the education process can really effect the increase of efficiency of production. This task has to be set as a obligatory target for projecting new equipments. An analysis of our institute done in several branches of GDR economy shows that we are still far away from giving such targets for all automation projects. We found out that this aspect of speeding up scientific-technical progress is still widely underestimated. That means that in management and planning of production processes the creation of working places where it is possible to utilize the increased qualification level does not yet play the role due to its importance.

We think that in order to fulfill the target of the five-year-plan directive for 1976 to 1980 about creation and reconstruction of at least 180,000 working places a year management and planning have especially to be concerned with following problems:

1. Already at the stage of projecting and constructing of new equipments targets have to be set to which degree unskilled and hard physical work can be reduced and what kind of qualification is needed. In this field there still are real sources for an efficiency increase. The exact determination of contents and extent of necessary qualification can essentially contribute to raise productivity.

2. Planning and realization of investments require a well-timed determination of the necessary abilities and knowledge for operating new equipments. The enterprises must be given programmes for additional training, as far as possible in a programmed form. Underestimation of that aspect of investment preparation leads to economic losses such as breakdowns, down time etc. with an amount of some hundred of millions of marks a year.

3. The high quota of women in unskilled labour still is a fact. Therefore everything has to be done to prevent that negative effects of differentiation of qualification requirements in connection with the scientific-technical progress only affect women. By means of scientific work organization everything has to be done in order to reduce such operations which require only simple and unskilled work as well as to reduce monotony of work and one-sided physical and psychic burden. It is necessary to analyse and generalize experiences made in organizing job rotation and in improving the mental climate within working teams. In cases where equipments still require one-sided and unskilled operations in a large scale possibilities have to be found out and realized for creating substiantal operations by applying new forms of combination of labour, of planned job rotation and of job enlargement.

4. The better utilization of the increased education and qualification level does not mean blocking such lines of scientific technical progress which are connected with a reduction of qualification requirements. Doing so would mean conserving mentally complicated technologies which are difficult to master.

So for instance qualification requirements for operation the first computer generation was incomparable higher than the present needed qualification requirements for operating equipments with microprocessors. Simplification of operating and reduction of the training period are without doubt an essential aspect of scientific-technical progress and substantially contribute to its rapid extension and to an increase of its economic efficiency. The reduction of mental requirements can of course have positive consequences for the development of personality. With that a considerable higher rate of working people is enabled to gain the necessary qualification for operating automatic equipments in shortest time and so to have knowledge on various fields. In order to emphasize the problem: the way we have to go is not conserving mentally difficult operations, far more, by means of a scientific work organization - which takes the increased qualification level into account - we have to give working people the possibility to use their physical, social and mental abilities in various activities in the process of increasing the efficiency of labour. Management and planning of production have to guarantee such a scientific work organization that brings into action modern technology as well as increased qualification level of working people for a growth of social labour productivity and efficiency.

It is necessary to consider the differing situations skilled workers have to manage in the various states of mechanization and automation. In mechanized production the qualification of a skilled worker is needed permanently for producing a special product. In this case qualification is a precondition for the machining operation. Though in automated production skilled worker qualification is an absolute precondition for production, it is not directly needed for carrying out the typical operation in automated production such as steering, regulating and controlling. So skilled worker qualification in automated production is not permanently used. In order to prevent a permanent undercharge of skilled workers scientific work organization has to find new kinds of division of labour and with this to enrich and enlarge the work.*

So scientific work organization has to gurantee:

- formation of new kinds of combination of labour for creating substiantal and ambitious operations
- increase of responsibility of working people in management and planning of production process
- planned job rotation
- possibilities of identification with the final product
- balance between physical and mental requirements
- job enlargement
- possibilities for communication and cooperation
- possibilities for realizing ideas and initiatives
- pleasant working environment.

An important component of socialist way of life is the permanent in-service training of working people. It sets important preconditions for an universal development of working people. At present in the GDR about 2 million people participate in both a systematic education and in-service training (among them 50% women). Without doubt a permanent in-service training will in future play an important role in the system of needs of a socialist society and in the further development of the socialist way of life. In the last years adult education essentially contributed to the increase of the number of skilled workers and of university and technical school graduates. 30 to 40% of today's skilled workers in the GDR got their qualification by adult education. Every third university graduate and every second technical school graduate got his diploma by way of external studies. Taking the now reached high qualification level into account adult education will in future take up its actual function, reproduction of existing qualification level (slide 10).

Under conditions of speeding up scientific-technical progress we are confronted with the obsolescence of knowledge. The dimension of necessary requirements to in-service training can be illustrated by the following consideration: Assuming a 3%-progress rate of knowledge you come to

---

* Annotation: E.M. Langen: Zum Zusammenhang zwischen der Entwicklung von Mechanisierung und Automatisierung und der Qualifikation der Produktionsarbeiter (Connection between mechanization/automation and qualification of production workers), Dissertation, Berlin 1977, S. 184.

following expenditure of time needed for in-service training during the whole period working life: skilled workers 2,8 years, technical school graduates 6,7 years, university graduates 7,8 years. That means 10% of the working life period of skilled workers, 24% of that of technical school graduates, 28% of that of university graduates (slide 11). Of course there is no strong law of an increasing expenditure of time for in-service training. But these figures show how important it is to take measures in time in order to meet the process of obsolescence of knowledge and to care for an interlacing of education and work processes so that in-service training will get an adequate place in the socialist way of life. All this emphasizes the importance of general education which sets the preconditions for a permanent interlacing of existing knowledge with new knowledge, for a permanent transformation of existing knowledge to the socially necessary level by in-service training.

The requirements resulting from the rate of progress of knowledge show that we need an effective combination between three components of in-service training:

- in-service training in an organized form

- in-service training according to own interest and initiatives

- in-service-training by an active participation in various forms of the intellectual-cultural life of socialist society.

Of course these three forms are mutually connected. They influence each other to a very high degree. The organized form of in-service training can only fulfill its actual task if it is based on the other two components and vice versa supports them effectively. At present the organized form of in-service training in the GDR is realized to about 80% during work hours. That does not only bring along high economic expenditure in direct and indirect forms but is furthermore connected with another danger: If in-service training is too tightly bound to the direct operations and tasks of a working place it cannot fulfull its stimulating function for participation in intellectual- cultural life. Therefore oranized in-service training has not only to give working people knowledge directly applicable to working tasks but also to contribute to the development of political and philosophical thinking as well as to the enrichment of intellectual and cultural life. At the same time new possibilities for the two other forms of in-service training have to be found by a higher efficiency of the spare time of working people.

slides 1 and 2

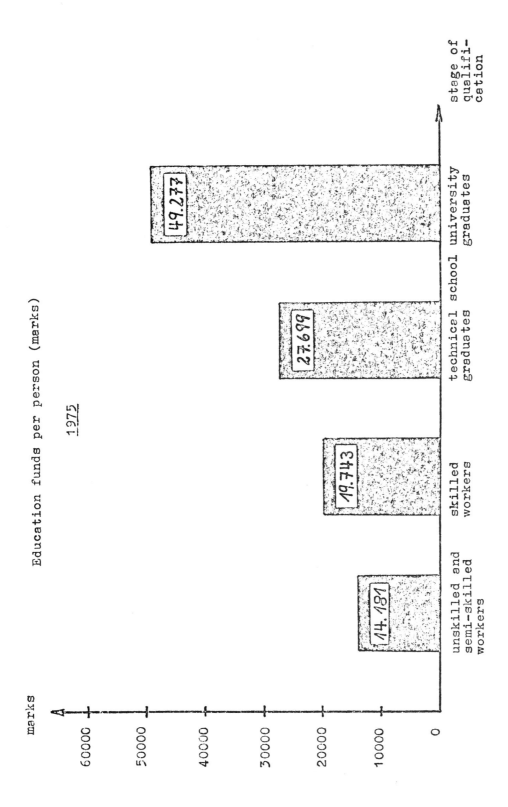

slide 3

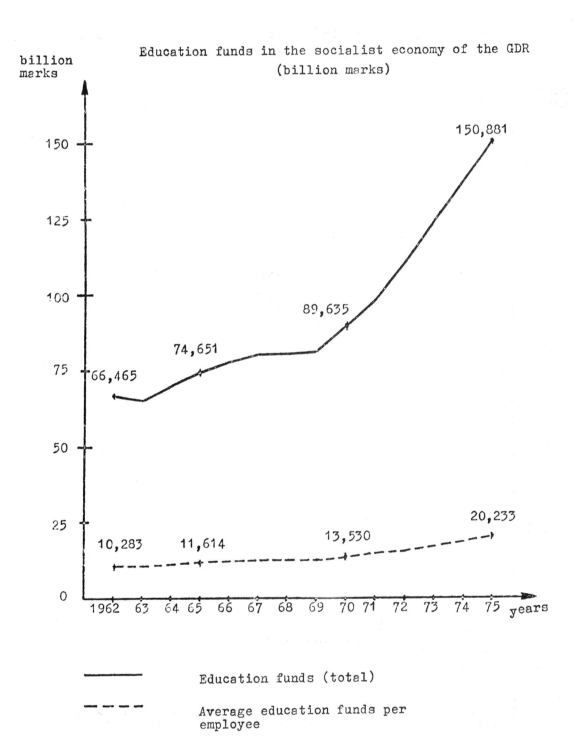

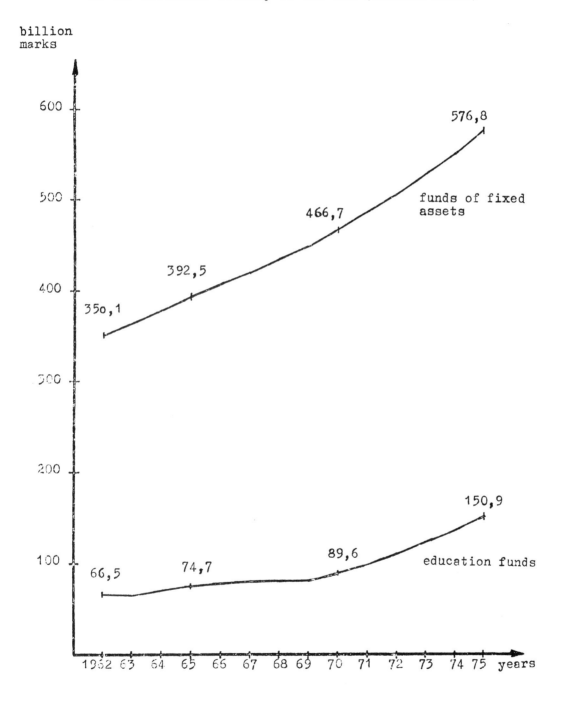

slide 5

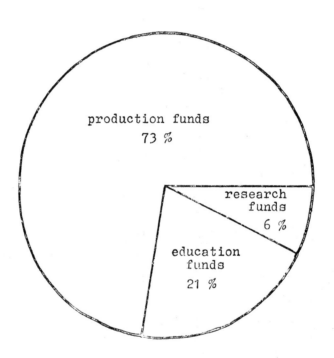

slide 5a

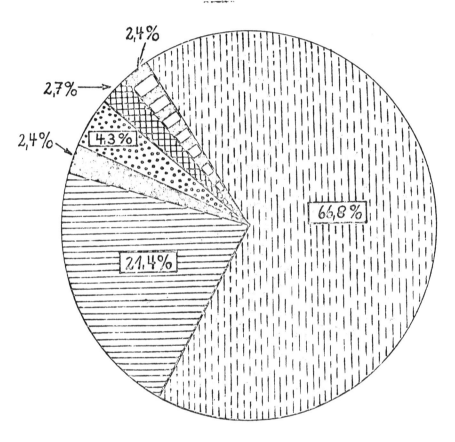

slide 6

Distribution of technical school graduates to branches of socialist economy of the GDR (percent)

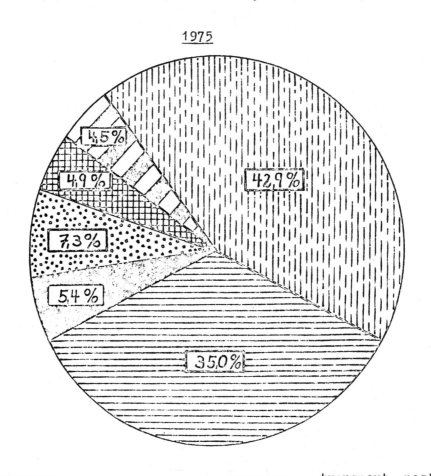

	industry		transport, posts and telegraphs
	building industry		trade
	agriculture		other producing and non-producing branches

slide 7

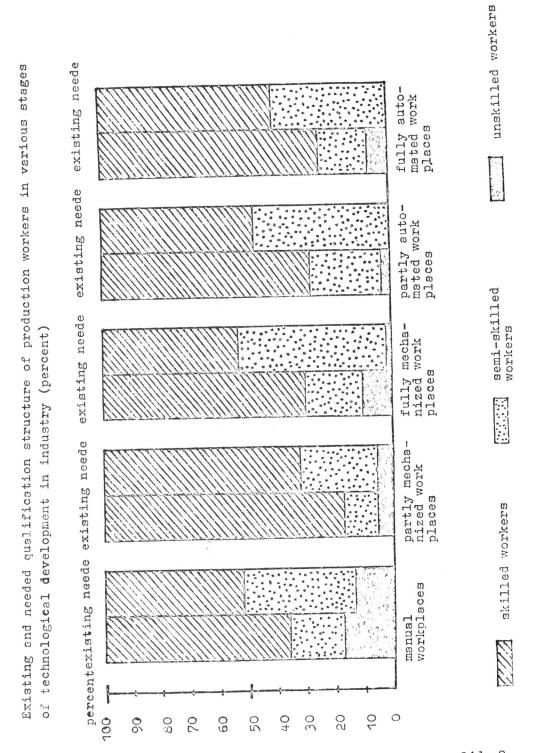

slide 8

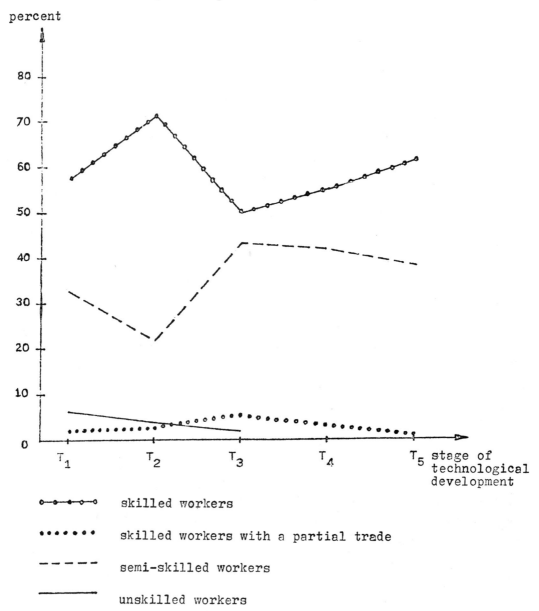

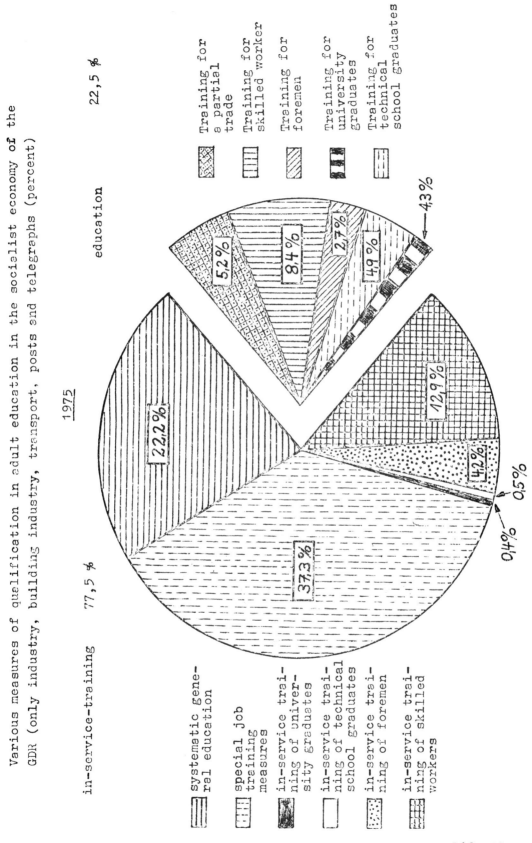

Various measures of qualification in adult education in the socialist economy of the GDR (only industry, building industry, transport, posts and telegraphs (percent)

slide 10

Full-time education, in-service training and period of working life of qualified employees

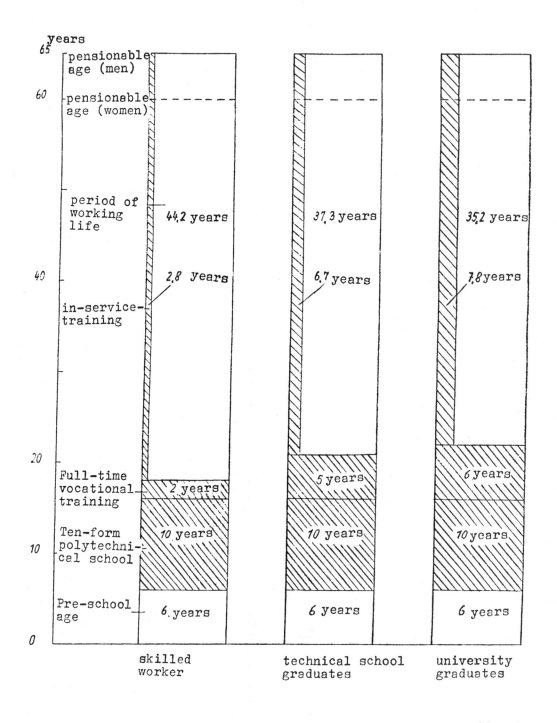

slide 11

THE SOCIO-POLITICAL RESPONSIBILITY OF CONTROL ENGINEERS FOR THE TECHNICAL DEVELOPMENT OF THE FUTURE

E. Welfonder and K. Henning*

*Department for Electrical Power Generation and Automation Technique,
University of Stuttgart, Germany*
*Institute for Automatic Control, Technical University of Aachen, Germany

1. INTRODUCTION

All of us that work in the field of automatic control - whether in research, teaching or application - are continuously trying to find new and better control concepts and looking for their suitable realization - today often with the aid of modern data-processing, such as process and micro-computers.

Besides all these professionally required efforts we are more and more confronted with the general question of where the technical development of the next few years or decades will be leading us (a question which is also concerning the control engineer).

For technical progress is accelerating at an increasing speed, that is: in shorter and shorter periods of time continuously improved new products have to be put into the market, and the engineer is expected day after day to devote all his energy to fulfilling the targets of this striving for progress.

2. ACCELERATION OF TECHNICAL DEVELOPMENT

This continuous acceleration of technical development can easily be seen from the tendencies shown <u>in figure 1</u>, referring to:

- global, cultural and technical development (furthermore see /1,2/)
- industrial development
- component development for data processing and automation technics.

Any engineer or practical scientist facing these facts clearly has to realize that this nearly exponential trend of development cannot continue for many decades. In this context also see studies of the Club of Rome about "The Limits of Growth" /3-6/.

Therein modern automation technics are having an essential share in the continuous acceleration of progress:
For by modern means of data processing the economical, developmental and productional running is being more and more optimized. The resulting rationalization measures are leading to a continuously decreasing demand for labour per production or activity unit, and this not only is the job sector, but also in the private sector as for instance by labour saving household machines, and therefore additionally housewives try to return to jobs again. As - in spite of the steadily decreasing demand for labour - full employment is still aimed at for ethical, social and also political reasons, there remains - apart from the limited possibilities of reducing working time - only one way out and that is an increase in production.

Therefore employers are - for reasons of industrial economics - continuously trying to find additional outlets for their products by means of opening new markets as well as by creating new demands. Therein the employers are making use of the new technical developments more and more quickly.

In the same way politicans are - because of their national economies - striving for a continuous economic growth. So the question of where there are the limits of the growth of production also arises here.

3. PRODUCTION OPTIMIZATION CRITERION

There is hence a great discrepancy between: on the one hand the unrestricted striving for progress of the individual producers as well as of the industrial nations in competition with each other and on the other hand the only limited possibilities for progress in the near future.

If one asks for the general reasons of this unrestricted striving for progress, the answer is in the first place to be seen in the still dominating production optimization criterion:

$$\text{Aim of production} = \text{economic profit}/_{\text{maximum}} \text{ material benefit}/_{\text{maximum}}$$

Due to this "one-sided" optimization criterion - regarding only the economic costs - employers, engineers and politicians let themselves be guided by a seeming eigen-dynamic of the technological development. They do so by putting into practice what can be technically done, as soon as it seems economically lucrative - without sufficiently considering socio-political restrictions necessary for our time as well as for the future, see <u>figure 2</u>.

In this context one has to take into account that legal provisions - such as anti-pollution laws or provisions for working conditions guaranteed by trade inspection autorities or by the trade unions - are normally only made after the maximum limit range of what is tolerable for the man-environment relationship has been surpassed strongly, see <u>figure 3</u>, curves a, b; furtheron the results of passed laws (curve c) lie very often only slightly below that tolerable maximum limit range and that even though quite often production-methods could be realized that cause considerably less damage (curve d).

Also professional associations such as trade unions and technical organization ions have not take a sufficient interest in these overall socio-political tasks. They still more emphasize problems inside of the factories and payment-problems as well as the documentation and standardization of technical knowledge respectively (see <u>figure 4</u>).

According to the above mentioned facts it is easily to be understood that the industrial enterprises did not begin to solve the increasing problems arising from technical effects - for instance the recleaning of waste water as well as the removal of unclear waste - in due time. For the arising costs are contradictory to predominantly economic optimization criterion.

Nor did the political authorities take care of the problem in time. Because of unsufficient knowledge they did not recognize the technical effects in due time.

Thus, during the last few years the citizens have grown aware of a lot of actual damage arising from technics but most of the citizens are not able to fully understand the technical correlations and therefore feel more and more insecure and don't see any way out.

Not seldom were the political authorities only informed by the newspapers about negative effects of new technical developments, and often the pressure of initiatives of the citizens was necessary to force the politicians to take the required legal measures.

Therefore it is highly unsatisfying that negative effects of technical developments are indeed often recognized by experts in time; whereas industrial enterprises, political authorities as well as professional associations, don't take the required steps in due time. To improve this situation a new production optimization criterion will be necessary, where besides the profitableness also the social compatibility has to be taken into consideration, so f.i. in the following form:

$$\text{aim and objects of production (now and in the foreseeable future)} = +k_M \begin{array}{l} \text{economic profit}/_{\text{max.}} \\ \text{material advantage}/_{\text{max.}} \\ \text{advantage for men}/_{\text{max.}} \\ \text{technical assesments on men}/_{\text{min.}} \\ \text{advantage for environment}/_{\text{max.}} \end{array} + k_U \begin{array}{l} \text{technical assessments on environment}/_{\text{min.}} \end{array}$$

Such a comprehensive optimization criterion regarding the full production costs:
- can be easily formulated in general form
- cannot be easily be provided with quantitative weighting-factors related to man-environment relationship
- and cannot realized by any means within a short period.

Therefore the main task of such a new production optimization criterion shall be to inspire a new thinking about technical progress
- being highly necessary from the socio-political point of view. Therein
- it should be at present taken into consideration by all those citizens who already have a sufficient sense of socio-political responsibility
- during the following years it should be a part of education and training in all sorts of professions
- and building up on this its realization within the enterprises as well as regional, national and international, should systematically be executed for years to come.

4. <u>SOCIO-POLITICAL RESPONSIBILITY OF ENGINEERS</u>

Due to the technical progress technology is becoming more and more complicated, and the technical realization will mostly be more complex.

Refined procedures on the one hand increase the quantity and quality of production, as well as the use and re-use of raw-materials and energy of low quality; on the other hand, however, refined procedures include the danger of undesired negative effects more strongly.

And more complex production installations where different technics and process components are working together can furtheron only be understood - at least their detailed effects - by a limited number of experts, mostly engineers and practical scientists.

Thus with the acceleration of the technical development engineers have to accept a growing socio-political responsibility.

Non-technicans (like most of the politicians, doctors, sociologists, technologians, lawyers, etc.) who work for men and society - for whom technics shall bring advantage - are mostly not able any more to see the possible dangers being caused by technics.

This discrepancy results from the fact, that for many and many years engineers regarded it as their task to design technical equipment. It was not their problem how the men - using this equipment - got along with it. Even today, the actual technical problems i.e. data bank systems, the use of nuclear power plants or the pricnciples of automation of the economy by means of economic planning methods are performed in a way that shows that the possibilities and dangers of the technical effects in a highly technicalized society are not fully understood.

The critical point of this development is, that the industrial nations - Eastern and Western - uncritically export their ideas of development into the third and fourth world; with the aim that the industrial nations shall survive.

Therefore engineers have to make all efforts to recognize the dangers and the damage arising from technics as early as possible. It is their responsibility to demonstrate these dangers and damages in clear language - understandable also for non-technicians - and to present constructive propositions for improvements, and this if possible already in the planning-phase.

5. INTERDISCIPLINARY COOPERATION- NECESSITY AND POSSIBILITIES

After having demonstrated the possibly dangerous effects of technics engineers have to discuss them with experts for human beings as well as with experts competent for the environment, see picture 5.

By such an interdisciplinary discussion it must be found out whether the possible technical dangers will cause damage and if so:

- of what kind
- to which extent
- and during which periods.

If dangers are to be expected, engineers have to make suitable propositions for improvement and to check them as to their practical use - within the sense of the extended production optimization criterion together with economic experts, competent officials, politicians etc.

Moreover, it has to be made sure that these improvements and measures will be realized also in case of future projects. The respective provisions have to be officially made according to existing or new laws.

6. PROPOSITIONS RELATING TO INTERDISCIPLINARY COOPERATION

Now we are coming to the question of how interdisciplinary cooperation can be realized in such a way that the individual engineer or scientist has a chance effectfully to show recognized dangers and damage due to technics. In the following we are mentioning two possibilities:

I. The same article concerning negative effects of technics will be published in various expert magazines - such as science, technics, medicin, biology, sociology, theology, etc. The comments will be equally published in all these magazines and thus, an interdisciplinary discussion will be initiated.

II. Existing expert associations or newly created institutions will take over the responsibility for the coordination of the interdisciplinary cooperation; i.e. they will have:

- to collect and to discuss interdisciplinary indications announced by engineers with regard to possible negative effects of technics and to arrange the interdisciplinary discussing of such indications in suitable ways
- to stimulate experts of the different faculties to cooperate
- to collect and, if necessary, to interpret the results of the interdisciplinary discussions
- to inform the competent official authorities of the propositions

for improvement and of all measures concerning dangers and damage of the technics

- to give also the politicians a chance to avail themselves of the prepared expert information concerning technical assessments as well as suitable counter-measures within a short time.

7. AIMS AND OBJECTS REGARDING THE FUTURE DEVELOPMENT OF AUTOMATION

In the precedent part the increasing responsibility of engineers in the socio-political field as well as the necessity of interdisciplinary cooperation has been pointed out. In the following there will be demonstrated which problems and questions the interdisciplinary expert teams have to deal with in the sphere of automation.

In order to give a clear survey, we are using the method of a "goal system" that besides other things is used to mark the development frame of enterprises and firms or the aim- and object-frame of associations (see for instance the goal system of the VDI /7/).

The suggested goal system regarding the future development of automation shall be the basis for discussion and does not claim to be perfect or complete. Rather shall a basis for discussion be given for the experts who analyse the social effects of automation in the widest sense. This basis shall on the one hand give a better survey of the number of problems and the various institutions that deal with these problems, and on the other hand form a basis for the specific work programme of the respective interdisciplinary team of experts.

The goal-programme - figure 6 - contains a basic guiding principle: the future development of automation must be adapted to the respective cultural situation, i.e. the extent and the kind of automation will to a high extent depend on the fact, whether the installation will be made in a highly industrialized country or in a developing country.

Under this point of view it must be tried to find chief points of view that have to be taken into consideration with regard to the future development of automation:

- humanisation of automation
- large- and small-scale automation
- human situation in man-machine-interface systems
- valuation of automation.

Further more, it must be discussed whether these four points include all essential components that are to be considered with regard to a culturally adapted development of automation in the future. For the individual spheres a so-called "goal-frame" can be fixed. In the case of humanization of automation i.e. the degree of employment must be found out (point 3.1.1.). In a similar way i.e. the limits of human abilities in relation to supervisory tasks have to be discussed (point 3.3.1.). As to the valuation of automation a goal frame concerning the effects on philosophy could be laid down.

Concrete programmes have to be formulated and carried out within the clearly defined goal frame. If there i.e. were any effects on philosophy, the effects of the degree of automation on the general valuation of technics must be investigated. On a last fifth level we must put the question who respectively which institution is working at which special task. As to the above mentioned example it should be pointed out to the central VDI-group in the Federal Republic of Germany. This VDI-group thoroughly analyses the problems arising from human beings and technics. In a similar way, a programme could be defined from the "goal-frame" effects of micro-computers on automation, a programme pointing out the social consequences arising from the application of micro-computers for new spheres. Also in this case it has to be cleared in a fifth level who will be delegated with these questions to the full extent.

For further discussion the individual spheres are once more divided up in the figures 7-10. For instance, regarding the field of humanization of automation you will find propositions for the goal frame:

- automation and employment
- working places
- automation and organization

In figure 7-10 you do not find a complete scheme of programmes and persons, institutions, etc., who are engaged in this problems. It is rather like a mosaic, where some places are already filled up. The reader may complete these figures with his own knowledge and interests.

It is obvious, that there never will be a complete scheme for the whole problem of a cultural adapted automation. Nevertheless one should try to structure these questions as far as possible.

8. RESUME

In this paper we have tried to point out the problems of future automation and its social context in connection with the general development of technics. The goal-system presented therein could be a basis for the discussion concerning the future work of the IFAC-Committee "Social Effects of Automation". It is therefore necessary to complete the goal-system especially with regard to the single programmes and to level 5 "Who does what"? This does not mean, however, that the topic must be comprehensively dealt with. Due to the modular structure of the system it is easily possible to keep - in addition to the committee-work - the general ideas for a future automation every time up-to-date. Extremely necessary will be the introduction of work teams, of special magazines and the cooperation of experts who dedicate their full strength to the respective goal-programmes. All the more, since during the last few years many group and institutions have started to deal with these problems. On the basis of a nearly complete goal-system a new special point may be found for the work of the committee "Social Effects of Automation" which may substitute or follow the previous goal-frame "working places" (point 3.1.2.)

It seems to be urgently required for the future work of the committee, on the one hand not to renounce the concrete work of the goal-frame level 4", but on the other hand to take into care careful consideration the whole complex of a culturally adapted automation, in a more intensive way than during the past. If this committee will duly fulfil its task to make, on an international level, a profession-based political contribution of control engineers - on the basis of an interdisciplinary team - such a concerted action of the committee is absolutely necessary.

9. LITERATURE

/1/ Eigen, M.: Das Spiel; Naturgesetze steuern den Zufall. Piper Verlag München, 1976.

/2/ Beer, R.: Mensch, Technik, Wissenschaft. Technische Rundschau Nr. 26, 28.6.1977.

/3/ Meadows, D.C.: The Limits of Growth. Universe Books, New York, 1972.

/4/ Mesarović, M. and E. Pestel: Menschheit am Wendepunkt. Deutsche Verlagsanstalt Stuttgart, 1974.

/5/ Gruhl, H.: Ein Planet wird geplündert. S. Fischer, Frankfurt, 1975

/6/ Alternatives to Growth 77. Second of Five Biennial World Conferences "The Nature of Growth in Equitable and Sustainable Societies." Houston, Texas, 2.-4.10. 1977.

/7/ VDI-Nachrichten, Dezember 1977.

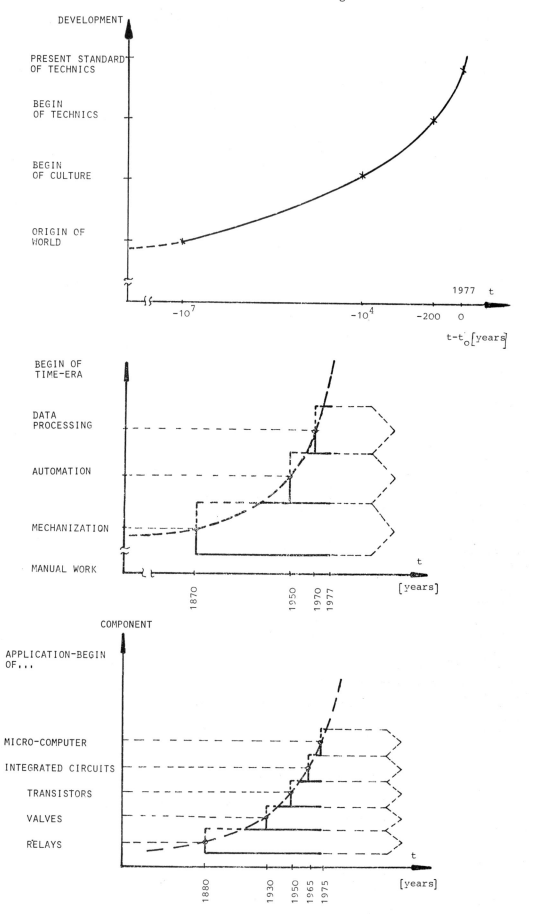

Figure 1: Tendencies in technical development

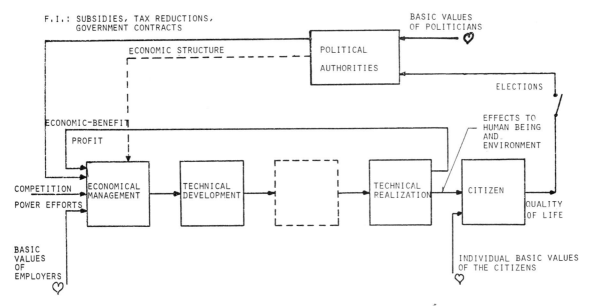

Figure 2: Technical development, mainly regarding economical interests

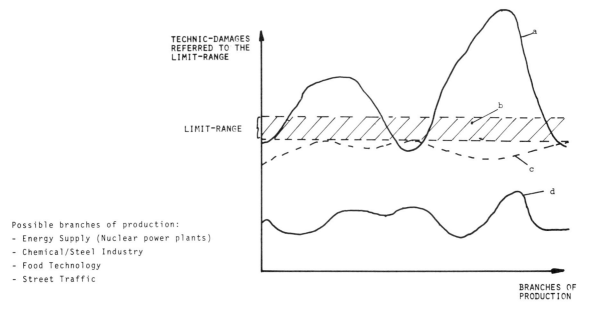

Possible branches of production:
- Energy Supply (Nuclear power plants)
- Chemical/Steel Industry
- Food Technology
- Street Traffic

a) Damage due to technics in case of exclusively profit optimization
b) Limits of damage caused by technics
c) Limitation of damage caused by technics by means of legal regulations
d) Damage caused by technics in case of an extended production optimization criterion (under consideration of the human being and environmental factors)

Figure 3: Negative effects of technics depending on the production optimization criterion

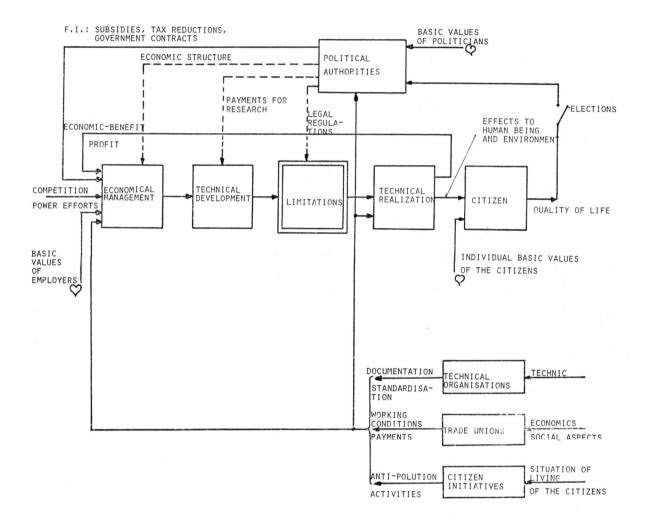

Figure 4: Influence on technical development of the political authorities and representations of special interests

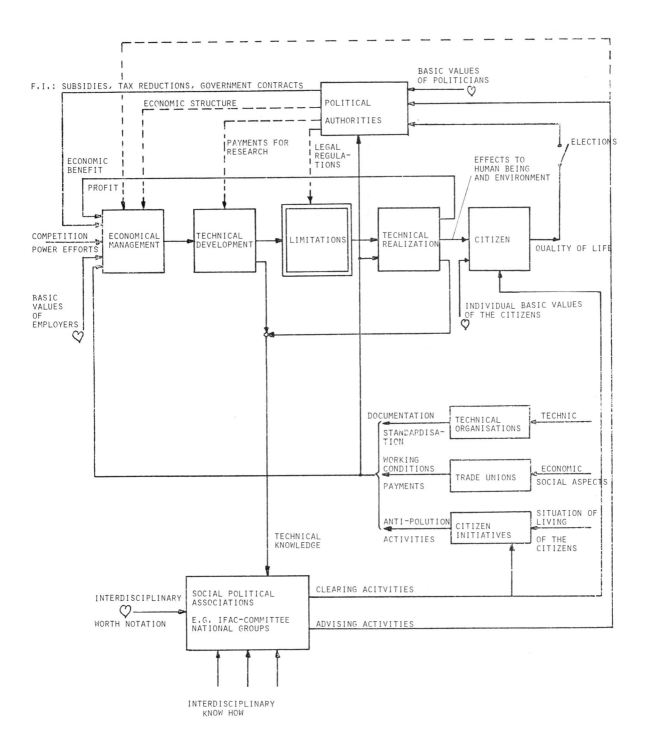

Figure 5: Influence on technical development by interdisciplinary groups

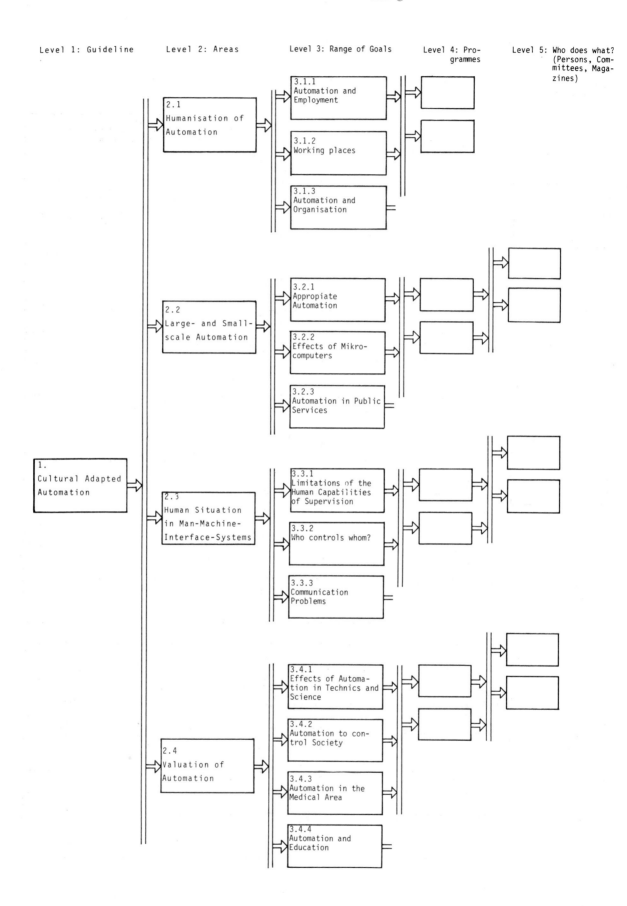

Figure 6: Guidelines for future development of automation

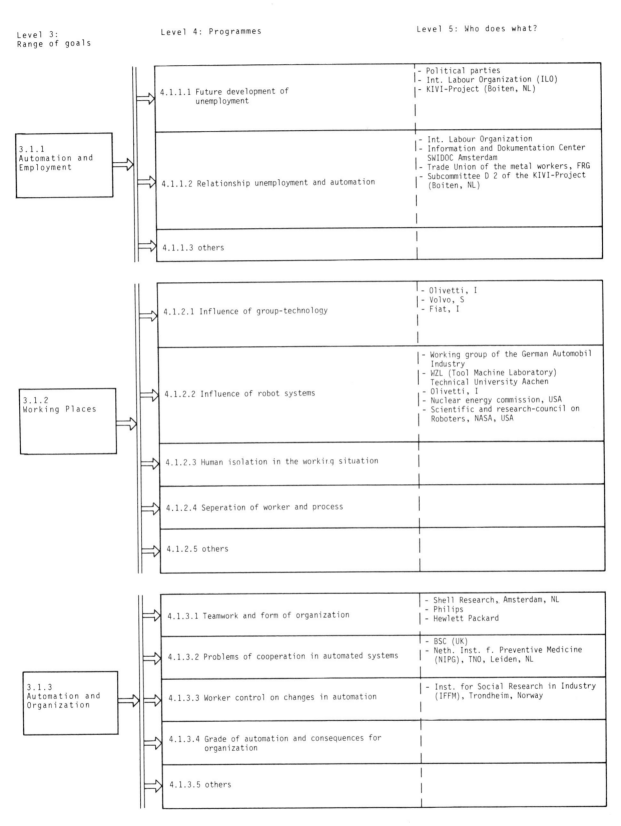

Figure 7: Humanisation of automation (Area 1)

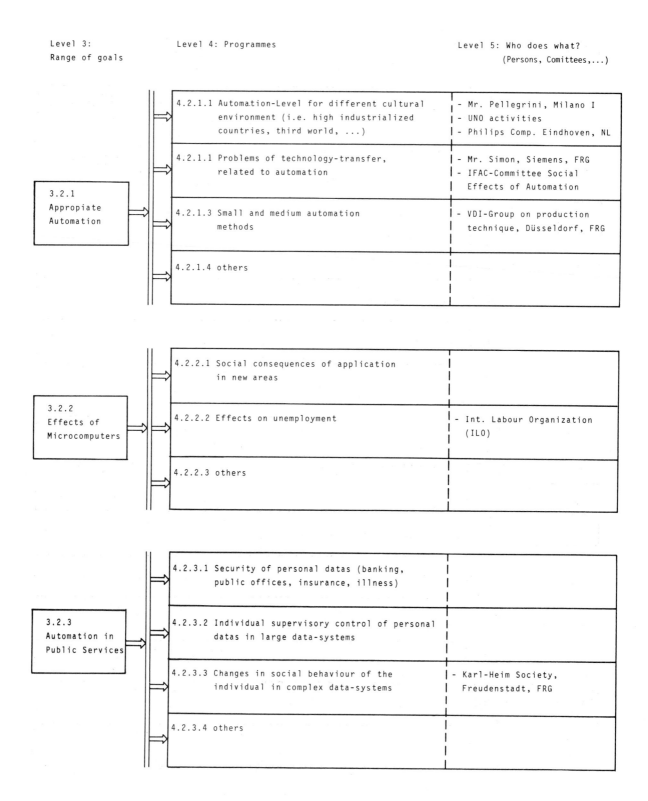

Figure 8: Large- and small-scale automation (area 2)

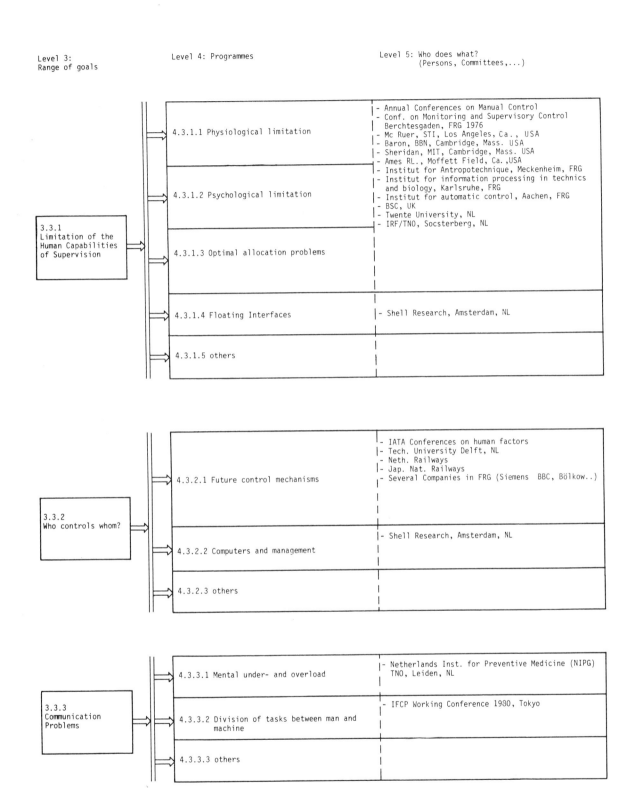

Figure 9: Human situation in man-machine-interface systems (area 3)

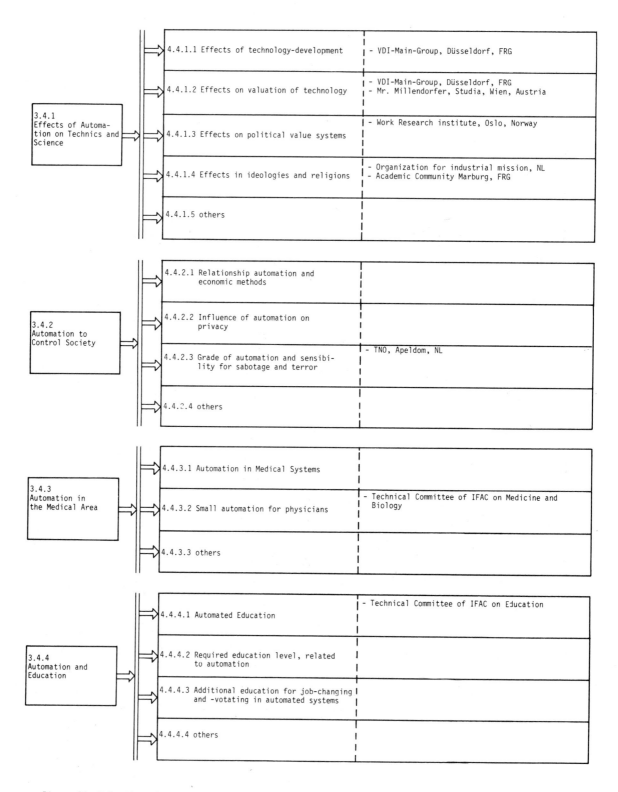

Figure 10: Valuation of automation (Area 4)

6. DISCUSSION REPORT

I. INTRODUCTORY SESSION

Chairman: **F. Margulies**
Co-chairman: **K. S. Bibby**

1. Words of welcome

Words of welcome were spoken by ir. H. Fekkers, who represented the Mayor of Enschede; prof.dr. I.W. van Spiegel, the Rector of Twente University of Technology; ir. Sj. Tysma, who represented the Division of Automatic Control of the Royal Institute of Engineers, the Dutch National Member Organization of IFAC.

2. The Design of Work: New Approaches and New Needs, E. Mumford

Enid Mumford started the presentation of her paper by comparing two models of the world: that held by engineers and that held by social scientists.
Engineering is presently centered on technical/rational ideas; the social sciences on social/emotional values. The problem is how to integrate these two models.
Social values are often considered as "soft", as opposed to the so-called "hard" technical values. However, this does not mean that they would be less important. People have strong personality needs (development of potential, esteem, security) and role needs (work interest, work challenge, work autonomy and work significance). Moreover, under favourable circumstances, people show interest in work; commitment to work; desire and opportunity to increase skill ; desire and opportunity to participate; and view work as a meaningful part of life.
There are two strategies for making progress in the design of work.
Humanization of work (through work place and job design) and
democratization of work (through a redistribution of power in the enterprise).
But there is not yet any change in the design of the machine system with these approaches.
Engineers should contribute to an integrated man-machine design in which the objective is high efficiency and high job satisfaction.
Machine systems should be designed in such a way that the operator is able to develop a skill; control the machine; work in a team; develop new methods.

In addition, conventional options should be provided for those who prefer not to have deeper involvement in their work.
In her conclusion, Enid Mumford posed the following questions for research:

- What are the job satisfaction needs of different groups?
- How flexible is existing technology*
  in human terms?
- How can new technology be made more flexible?
- Can automation be a liberator and not a tyrant?
- How can engineers predict the human consequences of their decisions?

<u>Discussion</u> (based on notes by Keith Bibby)

Some participants commented on the possibility for designing the machine to be an intelligent companion for the worker, and for the enhancement of worker/computer relationships through the humanization of software. The latter has become an economically viable objective owing to the ever decreasing cost of hardware.
Others indicated that engineers often do not know what social scientists mean by many of their concepts. Terms often seem to be fuzzy; effects, interpretations, and beliefs are rarely separated in a rigorous way, and too often, social scientists are satisfied with description, when prediction should be the real objective.

It was also suggested that the economic constraints upon change are often not recognized. The current economic situation and trends in Europe, coupled with fears about job

---

* Social scientists often use the term "technology" where engineers would speak about engineering.

security, also tend to make the work force resist experimental changes.
These factors operate against engineers and social scientists engaging in joint long-term research projects.
Reference was also made to the problems of value conflicts between work designers, and people in positions of power at the top of firms, coupled with the resistance at all levels towards new hierarchical forms of organization.
On the other hand, it was felt that the pressure of legislation was beginning to change the distribution of power, and to create a climate of opinion more favourably disposed towards work force participation in the longer term affairs of enterprises, and that this whole learning process must be expected to be slow, and not without mistakes and failures.

3. <u>Quality of Working Life</u>: Comments on Recent Publications in English (prof. A.T.M. Wilson †)

Tom Wilson commented on the fragmentation of knowledge, which is particularly apparent in the universities. "God did not divide up reality in the same way as university chairs". The existing reward systems do not encourage interdisciplinary work.
In practice, however, industrial enterprises have to reckon with many different interactions with their environment. Balancing all these factors requires some system of Social Accountancy.

<u>Discussion</u> (based on notes by Keith Bibby)

Tom Wilson's paper developed into interesting dialogue about a broad range of subjects.
All changes in enterprises were characterized as being extremely difficult to manage in a controlled fashion. A black/white scheme was rejected as an adequate framework for examining the very complex open system of the modern enterprise.

The balancing of forces between reaction, survival and change was depicted as being similar to the feat of the man playing eight handed poker, particularly in view of the fact that pressures upon senior managers are so diverse, and the ideas and frameworks, in terms of which to analyze their situations, are so inadequate. No more can issues be described in terms of the right of ownership, and challenges to that right, but we are now universally confronted with problems in the distribution of wealth, costs, and responsibilities. Our failure to correctly analyze and react to these problems is one of the major causes of the structural weakness of the post-industrial economies. The failure of old accounting methods had led to a new sense of "macro-economic modesty", characterized, for example, by the emergence of a hybrid journal entitled "Accounting, Organization and Society".

The artificiality of analyses couched purely in terms of the organization of work, and man's orientation to it, coupled with confusions about macro-organization, and macro-efficiency versus individual or group self-sufficiency as societal objectives all pointed to the fact that our knowledge and expertise concerning the interplay of economic, technological and cultural systems is woefully inadequate. Clearly, in the face of this challenge, slogans are no more an acceptable substitute for coherent initiatives to lift the constraints of technology upon our thinking and doing, and to create new structures within which human freedom can be supported.

Fred Margulies mentioned a guideline for the humanization of office buildings, which was based on answers to a questionaire sent by the Austrian Union of white-collar workers.

---

† Prof. Wilson died December 1978.

II. SUPERVISION OF AUTOMATED PROCESS

Chairman: **E. H. Boiten**
Co-chairman: **M. Levin**

1. Presentation of papers

The following papers were presented:

- Human Control Tasks, a Comparative Study in Different Man - Machine Systems, C.L. Ekkers, C.K. Pasmooij, A.A.F. Brouwers, and A. J. Janusch;

- System Development and Human Consequences in the Steel Industry, K.S. Bibby, G.N. Brander, and T.H. Penniall;

- Two cases from the Norwegian Process Industry, A.B. Aune;

- Jobs and VDU's, a Model Approach, S. Scholtens and A.J. Keja.

The papers consider two different aspects of automation and humanization of work.

Kees Ekkers c.s. and Arthur Aune dealt with the first aspect: social, psychological, and health effects of automation. These effects were discussed both at the level of work organization and at the level of society.

Keith Bibby, Schelto Scholtens and Arthur Aune discussed the second aspect: How to cope with these types of automation problems.
Keith Bibby also illustrated his presentation with some patterns of organization. Figure 1 shows four types of vertical relationships:
  Authoritarian Intervention;
  Authoritarian Delegation;
  Delegation Consultation;
  Consultative Delegation.

Figure 2 pictures "Management by Objectives", including personnel objectives.

Finally, Figure 3 indicates the various interactions in a production organization, centered around quality and output.

Fred Margulies remarked that these figures do not show the negotiations between unions and management/industrial relations.

Arthur Aune gave some more information about the effect of automation on the job structure in the PVC plant. The operators will normally be in the control room, and only go to the process equipment if they cannot take adequate action in the control room.

2. Discussion (based on notes by Morton Levin)

For process supervision jobs, underloading seems to be a more serious problem than overloading. Co de Jong mentioned that a possible way to avoid underloading in e.g. the oil refinery, is to let the operators "play" with the plant in order to generate new modes of operation. The final result is stored in the process computer.

Another topic of discussion was the isolated jobs created by automation, which conflicts with human social needs. It was remarked that most new workplaces are only suitable for introverts, while many are peopled by extraverts. Kees Ekkers c.s. have been looking for correlations between these types of personality factors and output variables, such as health and job satisfaction. However, no significant correlation was found. In fact, one should be very careful with distinctions such as introvert - extravert as they are based on a static view of personality and underestimate the influence of the environment on human characteristics.
It has been shown in many studies on human relations at work, that most people prefer to work within a group, or, at least with another person. The human aspects of work are experienced to be just as important as salary. Hence work places should be designed in such a way that social contacts are easily possible.

There was also a broad discussion about worker participation in the design and operation of man-machine systems. Worker participation should not only affect the lowest levels of the organization, but it should also have an influence on the higher levels. In practice, however, it is

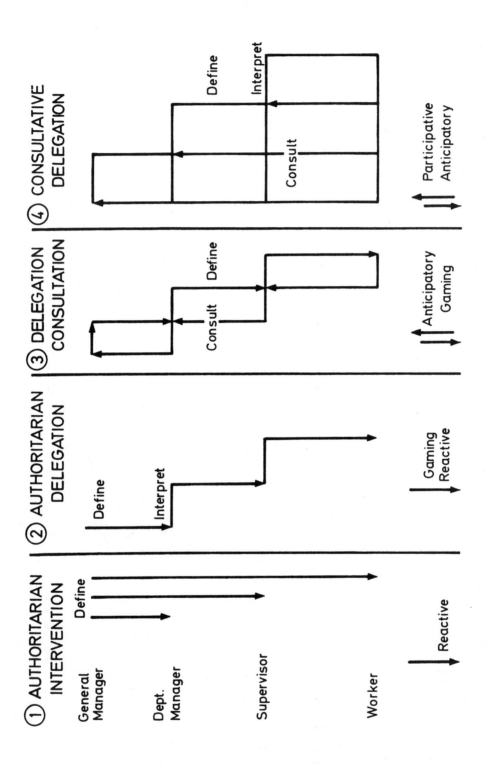

Figure 1: Organisational Relationships

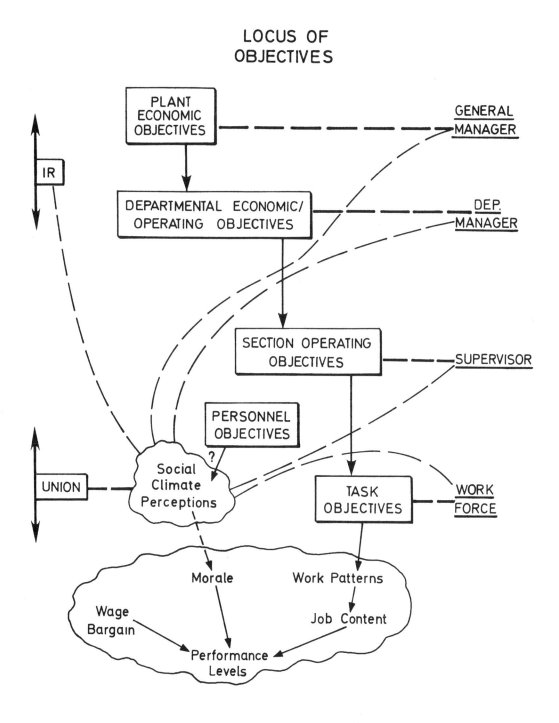

Figure 2: Management by objectives

not so easily realized. A difficult problem is the position of middle management; their work content should not be impoverished by e.g., job enrichment at the worker's level. Humanization of work must take place within certain limits, which are of technical, structural and economical nature. Action research appears to be a viable approach. It is characterized by a continuing dialogue between workers (and their representatives), managers, and researchers during the analysis of the situation and the working out of solutions.

It is very desirable that humanization of work is introduced right at the start of new projects. A problem could be the many other interests competing for management's attention. Management should therefore provide an organizational structure for taking everybody's contributions into account.

If we aim at fitting work to man, we have to envisage an alternative organization, together with an alternative technology. The alternative organization should give full opportunities for worker participation.

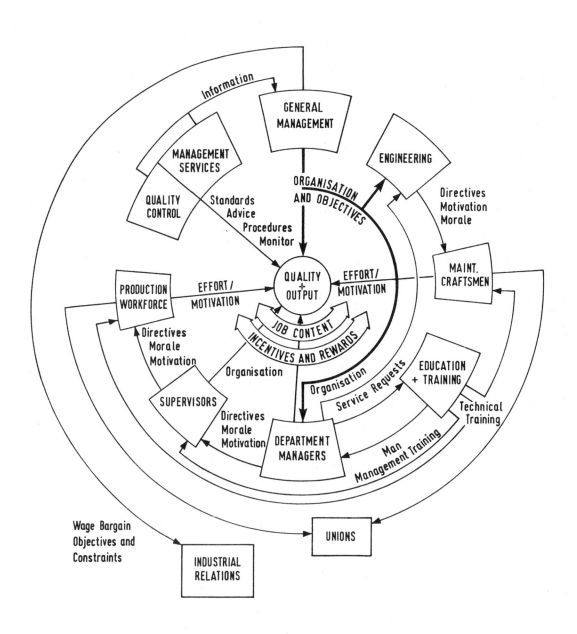

Figure 3: Interactions in a Production Organization

III. WORK ORGANIZATION

Chairman: **H. G. Stassen**
Co-chairman: **G. Wieser**

1. Presentation of papers

The following papers were presented:
- Basic Transformations: the Key Points in the Production Process, C. Schumacher (presented by George Brander);
- Distributed Cellular Manufacturing System, A.B. Aune;
- Automation and Humanization at the Dutch Railways; some Examples, D.P. Rookmaaker;
- Automation and Work Organization; an Interaction for Humanization of Work, J. Forslin.

Jan Forslin presented a nice sequence of diagrams to illustrate the historical development of work organization (see Appendix to this chapter).

2. Discussion (based on notes by Lena Mårtensson)

It is not necessarily so that the basic transformations in the production process, as defined by Schumacher, pose the most complex and demanding jobs for the workers. Other activities, such as transportation, and work piece scheduling in the steel industry, can be very demanding too. In many production processes, jobs can be enriched by making the workers responsible for programming (machine shops) or scheduling (automated processes).

A question was put to Arthur Aune about the kind of decentralization proposed in his paper. With the trend towards a post-industrial society (there is a prediction that only 5% of the people will work in industry in the USA in 2000), is decentralization still necessary? Arthur Aune pointed out the particular situation in Norway, where employment in fishing decreases, and oil production does not create many new jobs.

The social scientist's and the engineer's view on how the work is done is not the the same as that of the worker. The latter usually has no idea about theories, but of course he/she knows the work. It is important to find a common frame of references.
Hans Müller remarked that during the design of a new plant the workers are still in the old plant, hence it is difficult to obtain their contributions.
Training with computer systems can only be done on the job.
Moreover, continuous learning is very important after start-up of the new plant, for which financial means should be available.
When the worker has been able to evaluate different solutions, acceptance of the new system is enhanced.
Max Elden remarked that in Norway the law requires worker participation in planning. Actually people are put on the payroll for this activity. This is not always easy: A worker on the board of directors can feel like a hostage, as he obtains information that he cannot transmit to the workers.
Jim Crawley indicates two sets of forces: the work force to be involved, and the economic force: the cost of improving work environment and working conditions. Social scientists usually do not make predictions of costs and benefits, but unions and managements can be faced with, for instance, a conflict between number of jobs and quality of the working environment.

When changing the technology there is always an opportunity for changing the work organization. In some cases, this opportunity has been carefully considered, and the effect has been a meaningful and rewarding task. In other cases, however, this has not been done, which has led to more boring jobs than before the change.
The starting point for projects on humanization of work is not a matter of knowledge or scientific expertise, but a matter of values.
The task of the social sciences is a mobilizing one: To help people help themselves, to show the possibilities.

APPENDIX

AUTOMATION AND WORK ORGANIZATION
A THEORETICAL BACKGROUND

J. Forslin

Both studies I am going to present here show that there are possibilities in connection with a technological change in industry also to change the human qualities of work. In both cases technology has been designed to meet human requirements of a pleasant and healthy physical work environment. But only in one case has there been conscious efforts to improve the psychological conditions of work and this has mainly been done by means of work organization.

As a background for the two cases I would like to put the development they represent in a historical perspective.

Work organization has become a central issue during the 70's. Until recently there has been little communication between technical and social scientists working on industrial problems. There have been two quite different cultures, two languages and maybe two sets of values. But now it seems as if the field of work organization is an area where the two disciplines can meet and where each can contribute in a way that is meaningful to the other. You could say that work organization is the interlock between the social and the technical subsystems of a production system.

social system ⟨⟩ technical system
work organization

This area is meaningfully related to both the technical and the social side, and here each can apply its knowledge and find tools for action and change.

Why then has work organization become so important on a conscious level now and not earlier, for example when the big changes in work organization philosophy was brought about, e.g. at the introduction of MTM-systems in the 50's? In answer to this question I would like to stress the historical perspective on the development of production systems.

In my view there are three production forces:
- capital
- management
- labour

In the pre-industrial production all three forces were by and large united in the same person. The artisan provided the little capital that was needed for his primitive tools, he had all the knowledge needed about the production process and the market and also carried out the physical work.

capital ⟨⟩ labour
management

With the emergence of capitalistic industry there was a separation of capital and managerial functions from labour. The owner of production means was also the technical and administrative manager and had the contacts with the market.

capital management labour

When industrial units grew larger more capital was needed and the managerial processes increased in complexity. Ownership then became distributed on many persons, banks etc. and a professional managerial staff emerged and the three production forces were not entirely separated into different groups of persons.

capital management labour

One important consequence of this development has been dehumanization of work. That is, intellectual capacities such as creativity, initiative, problem solving has become unnecessary qualities for the majority of the industrial workers. What is demanded is sheer endurance. Man has become reduced to a physiological machine - by means of work organization, supported by technology.

The problems emerging out of this picture are both exploitation and alienation of the worker. In socialist countries - and elsewhere one may add - the focus has been on the conflict between labour and capital. The efforts have been aiming at in one way or the other to reunite these two production forces - but still leaving management of production outside. Not even in the self-managing Yugoslav system has the management of industries become the task of the workers.

capital labour

management

This might solve the problem of exploitation but not that of alienation.

In the west the recent emphasis is on reuniting management and labour to avoid alienation. Prosperity has so far in many countries decreased the problem of exploitation leaving capital untouched.

management labour

capital

Finally we now witness the emergence of various models to add also the capital into the unity - but on a collective level, e.g. employee investment funds. The circle would then be closed on a higher level.

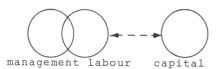
management labour   capital

Now - the separation of production forces happened a long time ago. This development towards alienation reached its peak in the industrialized world in the 50's. Why has it become an industrial issue now, being subject to legislation, fierce trade union attack, and protests from the workers in terms of labour turnover and absenteeism? A supplementary perspective on the development is needed here in order to fully understand the changes that industry is undergoing today.

This perspective is psychological and puts the focus on the change in the overt needs of the labour force that has been brought about by the prosperity of western society. If we in Maslowian tradition divide human needs into two categories we can describe the human set of needs as follows:

growth needs: - need for self-actualization
              - need for self-esteem
              - social needs

deficiency   -   "   "
needs   : - security needs
              - physiological needs.

As the needs have a hierarchical relationship it is implied that more basic needs have to be satisfied before higher needs become overt drives for human behaviour.

This gives us a tool to understand why some people are satisfied at the assembly line and others not. It also explains why younger generations refuse to take on industrial work even in the case of unemployment. Not until now have the basic needs of food, comfort and security become satisfactorily met thus exposing new latent needs for respected, meaningful, stimulating and developing work also in the big groups of workers. At the same time organization of work has moved in just the opposite direction creating less and less satisfying jobs though relieving man from the heavy and unhealthy tasks by means of technology.

What is happening now is experimentation in reorganizing work to regain its human qualities.

IV. EDUCATION AND TRAINING

Chairman: K. Henning
Co-chairman: B. Wahlström

There was one paper in this session:
Mechanization/automation and the development of the level of education and qualifications of GDR employees, by H. Maier.

During his presentation Harry Maier raised the question if nonsatisfaction should actually be encouraged to enable satisfaction by introducing an interest for change.
The stages of technological development indicated in slide 9 are:
- T1 : manual
- T2 : semi-mechanized
- T3 : fully mechanized
- T4 : semi-automated
- T5 : fully automated

Discussion (based on notes by Björn Wahlström)

It is very important that young people have the possibility to learn a profession. In the GDR 85% are using this possibility.
Fred Margulies brought up the importance of people obtaining jobs in accordance with their education and asked about hidden unemployment in the GDR.
Harry Maier answered that 25% of the educated and skilled workers are employed below their qualification, which can be interpreted as a positive interest in learning without demand.
In fact, the long term policy of the government is to use the creative potential of the people.
The goal for the end of the century is 25% university and technical school-trained; 65% skilled workers and 10% nonskilled ones.

VISIT TO N.V. PHILIPS' GLOEILAMPENFABRIEKEN, EINDHOVEN, NETHERLANDS

The participants paid a visit to Philips' factories at Eindhoven on Wednesday, 2 November 1977.
Mr. G.P. Mollerus (Public Relations Dept.) gave some information about the history and the organization of the company. This was followed by an introduction by Ir. H. Gelling (Social Affairs Department) about "Humanization of Work" in which he stressed recognition of the great learning abilities of people also on the shop floor.
After lunch, visits were arranged to the studio for electro-acoustics (Mr. A. Lammerée) and to the Mechanical Workshop, building SEU (Dr.ir. B. van Leusden).
More information about the latter can be found in "Work Structuring as a Criterion in Organizarion Renewal", Philip Wester, Social Research, Philips' Eindhoven (Netherlands) November 1977.

V. MANUFACTURING AND ROBOT TECHNOLOGY

Chairman: **J. Dockstader**
Co-chairman: **F. Muller**

Papers and presentation

(based on notes by François Muller)

- R.M.J. Withers and J.E. Rijnsdorp, Work of the Social Effects of Automation Committee from Bad Boll to Enschede.

  The session started with a brief presentation about past activities and current program of the IFAC committee on Social Effects of Automation.

- M. Misul, Work Organization with Multi-Purpose Assembly Robots.

  The report analyses the specific problems arising when introducing robot technology in a production process based on group technology. As the system was specifically developed "in house" for eliminating the existing bottlenecks of the production process, cooperation with the workers was easily realized and resulted in a functional layout acceptable to the work force. In particular it made the process more flexible with respect to time scheduling and production planning. As a result, thirty five of the assembly systems are presently in operation.
  A problem still is the feeding of parts to the robot, which cannot so easily be automated, and results in human tasks with poor content.

- L. Nemes, Man Machine Interfaces in the CONY - 16 Integrated Manufacturing system.

  This report emphasizes the need for considering data processing and information flow as a full-fledged production factor, and the necessity to use the acquired information as a feedback for optimizing the manufacturing system. However, in discrete manufacturing systems human interaction will not be superseded in the near future with the available techniques. It should be geared towards a higher level, i.e. the required information modifications (for operators) and the required communication links (for supervision). The main problems for realizing this aim are linked with information screening, information selection, and the choice of the proper input-output facilities.

- U. Lübbert, Automation of Wig-Welding.

  This report deals with a novel method of providing the welder with optical control techniques for high quality welding, thus eliminating the sheer physical stress of such operations. However, it has to be considered as an intermediate step towards fully robotized welding technology and implies a reorientation of the welder's role in manufacturing.

General discussion

(based on notes by François Muller, Tom Whiston and John Rijnsdorp)

Keith Bibby made a comment on Fig. 7 of Mario Misul's paper, where the workers are shown in rows, hence social contacts are not easy. The answer was that lay-out indeed is important. Moreover, buffers create freedom for individual workers to leave their work places from time to time.

Mario Misul also told something about the history of U.M.I. It was started in 1971 - 1973 with a pilot group of 25 workers. The attitude of the union initially was rather critical, but this changed with the growth in technical knowledge, and due to the fact that nobody was fired because of the new manufacturing approach.
The opinion of the workers was gathered in continuing, direct informal contacts, rather than by means of questionaires and the like.

László Nemes mentioned that the workers enjoyed the dialogue with the computer.

In commenting on the welder's job, Fred Margulies stressed that devaluating a skilled job is undesirable. Can we talk about humanization of work if interesting tasks are removed without adding new content?

Remmerswaal gave the opinion that welding is a very unsuitable task for human beings. Hence the only satisfactory solution is to eliminate it completely by automation.

Fred Margulies suggested to do away with specialized jobs, which easily become obsolete. People should be in a position where they can do various tasks, which can be promoted by general education and training.

On the allocation of tasks between man and computer, Co de Jong remarked that some tasks belong clearly on the side of the computer (e.g. multicascade control systems), while other tasks belong on the side of the operator (e.g. error diagnosing). For a third group of tasks, operator and computer both are effective (e.g. local optimization, starting-up, generation of process operating modes). Here a "floating interface" can be applied, which leaves the choice of computer control versus manual control to the operator, according to the changing work load at any given time.

Various contributions stressed that with an increasing degree of automation the average level of humanization of work also increases, however, this does not proceed steadily, but with its inherent ups and downs.

John Rijnsdorp indicated the humanization problems in the initial phases of a new technology, when so much attention is paid to technical aspects that human factors do not receive sufficient consideration.

Tom Whiston emphasized that humanization of work has to be placed in a wider policy context relating to educational and unemployment problems; which would probably be accentuated in the future. This is summarized in the diagram below:

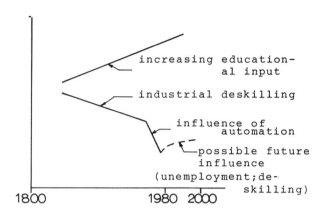

This diagram shows the massive deskilling which dates back to the first onset of large scale production procedures. Mechanized or partially automated procedures are at present accentuating the problem, and in this sense any moves to humanize work are important. However, a far greater problem relates to the future generation of alternative labour intensive and intellectually demanding tasks, which are pertinent to future unemployment patters and increasing societal education inputs. Tom Whiston also outlined very briefly some macro-strategies open to governments.

Finally, François Muller remarked that the company should continue to make profit in order to keep employment on the same level. If the production rate cannot be increased production efficiency should be improved.

VI. AUTOMOBILE ASSEMBLY

Chairman: **A. B. Aune**
Co-chairman: **J. Landeweerd**

Papers and Films

K. Namiki, H. Koga, S. Aida, and N. Honda.
An approach to the Production Line of Automobiles by Man-Computer Systems (presented by H. Koga)

- R. Knoll, Freeing the Operator from the Machine, as Exemplified on Assembly Lines in the Automotive Industry.

  Mr. Knoll illustrated his presentation by a film, showing the actual changes in production methods in the factory.

- H. Müller showed a film about the Fiat factory at Mirafiori (Italy), where the same robot carriers are being used which were also introduced in the Volvo factory, Kalmar (Sweden).

  In Mirafiori, the working groups are larger than at Kalmar. Autonomy of work is realized by large buffers: about 45 cars, which corresponds to one hour of average production.

General discussion

(based on notes by Lena Mårtensson and John Rijnsdorp)

Mr Knoll mentioned that the serial production line shown in the film has been replaced by a parallel arrangement, which offers more flexibility to the workers. In his opinion, the automation techniques as used by Volvo (Kalmar) and Fiat (Mirafiori) are more expensive, more space consuming, and give more difficulties with production of a mixture of car models.

In the case of Fiat, the increased investment costs have been balanced by labour savings and less absenteeism. Humanization of work should be efficient.

The measures taken at Daimler-Benz have improved product quality, as the worker now has the possibility to inspect and individually sign for his work.

There is also a decrease in scrap and recycling.

In the German Federal Republic, there is a shortage of skilled labour. The reaction of the workers and their acceptance of the changes have not been explicitedly investigated. In the case of Fiat, the previous situation was bad (crowded, heavy physical load), which favours acceptance of the new situation.

The acceptance is more favourable with skilled workers than with unskilled ones: The latter are mostly foreign workers, who only come to earn money during a number of years.

Fiat and Daimler-Benz are more interested in modifying existing factories (e.g. in view of space problems) than Volvo. Although all go into the direction of smaller factories, with Volvo the aim goes further (about 500 workers) than with e.g. Fiat (about 5000 workers).

VII. APPROPRIATE TECHNOLOGY

Chairman: **K. Henning**
Co-chairman: **B. Wahlström**

Paper

- A. Dell'Oro, U. Pellegrini, C. Roveda, Search of an Appropiate Transfer of Technology to a Developing Region within an Industrialized Country (presented by C. Roveda).

General discussion

François Muller pointed out that transfer of technology should not be restricted to production technology, but should also encompass utilization technology, e.g. of irrigation equipment.

Tom Wilson mentioned that the U.N. have indicated many examples of how not to do it. Positive results are relatively scarce thus far.

Fred Margulies stressed that the external expert cannot solve people's problems. One can only provide for better conditions so that they can solve the problems themselves.

VIII. SUMMING-UP AND FUTURE

Chairman: **J. J. de Jong**

Co-chairman: **Mrs M. E. Wilson**

Papers and presentations
- J.E. Rijnsdorp, Summing-up of the Workshop
- E. Welfonder and K. Henning, The Socio-Political Responsibility of Control Engineers for the Technical Development of the Future.

General discussion
(summarized by Mary Wilson and Co de Jong)

The discussion took off from the final section of John Rijnsdorp's paper. Looking back at the last workshop of this kind at Bad Boll some years ago, he suggested that the papers and discussions at that time showed a number of characteristics in the fields of automation and control engineering which were slowly being rectified:

1. In general, there had been too little consideration of human factors;
2. Insufficient recognition of the wide options and opportunities offered by new technologies;
3. Inadequate recognition of the constructive contribution which new designs of job and work organization could make to socio-technical production systems.

A wider conclusion emerged from these three points: the need, in general, to optimize man/machine relationships in industry.

John Dockstader summarized data and ideas from his study of the future, and his general conclusion was that it would be necessary to revamp all our societal institutions if there was to be a satisfactory future.

Fred Margulies reminded us of the need to remember the casualties and costs of change.

In the general discussion which followed - on the need to provide for variability and choice in terms of both products and the range of available jobs in industry - a number of conflict areas were indicated:

- conflict between technological achievement, economic and market possibilities and human acceptability
- conflict between the performance characteristics of human beings and of equipment
- conflict between the approaches of the natural, the engineering and the social sciences.

It was repeatedly pointed out that new types of organization are required to accomodate these conflicts to the greatest possible extent. This will require considerable adaptation from all social partners - management, the individual worker and the trade unions. Within an organization an effective two-way communication has to be developed. Fundamental problem areas are the following:

- strategic difficulty to introduce change because of psychological and cultural effects (human inertia, fear to lose established rights and reluctance to run risks which are difficult to predict)
- tactical difficulty because of lack of education, training and communication, capability and fear of delegation and of loss of efficiency.

Various forms of operator/computer interaction revealed in which direction present development is taking place. The following problem areas were indicated:

- economic limitations to experiments
- deskilling of the workers
- unemployment, also because intermediate levels in the organization are cut out
- victimization of the individual
- distribution of the available work
- work ethics
- position of the middle-manager.

In discussion of the paper by Welfonder and Henning, two points in particular attracted attention:

1. Consideration of the new concepts and evaluation of human factors in industry still tended to come too late in the planning process.

2. There was insufficient dialogue and inadequate co-operation between control engineers and social scientists - about human affairs in general, and about the needs and potential contributions of employees of all kinds.

Discussion of the first of these two points, the nature and timing of consideration of human factors in industrial planning, led on to a number of issues - for example, the importance of a full understanding of the nature, potential and limitations of these human factors; the significance of inadequate skill level in operative jobs to provide for those who whished it a degree of challenge and of the job satisfaction associated with this; the significance for all levels of organization of socio-technical change; and the emerging need to have regard to environmental factors.

In some concluding remarks, Fred Margulies, Chairman of the I.F.A.C. sponsoring committee, returned to the initial point made by John Rijnsdorp - the progress made since the Bad Boll workshop - progress in recognizing the significance of human factors and the joint responsibility of the social scientist and the engineer in increasing and maintaining this awareness. Both workshops had been concerned with identifying problems and promoting activities to tackle them. A first consideration in the humanization of work was the provision of the opportunity to earn and work for this purpose. In his view, the history of the last thirty years did not show a clear and high correlation between rapid development of technology and growth of unemployment. Where high unemployment occurs, the fault must be sought in the economic system and not in the production system.

There was great need for technologists to cross borderlines and make contact not only with social scientists but also with economists. Humanization involved maximization of flexibility in the use of technology and a great improvement in the range of factors taken into account in the planning and control of industrial processes.

Finally, he called for votes of thanks which were given with acclamation, to the Twente University, to Professor Rijnsdorp, Leo Verhagen and all enthusiastic staff members; to the presenters of papers, the discussants and all others who contributed to the success of the workshop. He also mentioned that he will retire as chairman of the I.F.A.C. Committee in 1978, but he had found Professor Rijnsdorp willing and prepared to take over this position from him.

LIST OF PARTICIPANTS

IFAC. Workshop on Case Studies in Automation, related to Humanisation of Work

NAME	ADDRESS	COUNTRY
Aune, A.B.	SINTEF, Div. Autom. Control 7034 Trondheim - NTH	N
Bibby, K.S. Brander, G	British Steel Corporation 140 Battersea Park Road London SW11 4LZ	UK
Bisterbosch, H	Emmastraat 55, Enschede	NL
Blum, U	IG Metall Frankfurt/Main Wilhelm Leuschnerstr. 79	FRG
Boiten, E.H.	Hoogovens, IJmuiden	NL
Boon, L.A.	Hoogovens, IJmuiden	NL
Brödner, P.	DFVLR, Kölnerstr. 64 Bonn-Bad Godesberg	FRG
Brouwers, A.A.	N.I.P.G., Postbus 124, Leiden	NL
Crawley, J.	British Steel Corporation 140 Battersea Park Road London SW11 4LZ	UK
Dockstader, J.H.	Hewlett Packard, 175, Wyman Street, Waltham MA 02154	USA
Dubel, J.	UKF - IJmuiden	NL
Elden, M	IFIM, Trondheim - NTH	N
Ekkers, C.C.	N.I.P.G., Postbus 124, Leiden	NL
Foeken, H.J.	T.H.E., Bedrijfskunde, Postbus 513, Eindhoven	NL
Forslin, J.	Swedish Council for Personnel Administration, Stockholm	S
Groeneveld, H.M.	Spanjaardsplein 16, Borne	NL
Henning, K.	R.W.T.H., Aachen Templergraben	FRG
De Jong, J.J.	KSLA, Postbus 3003, Amsterdam	NL
Kalsbeek, J.	Afdeling EL, T.H.T., Postbus 217, Enschede	NL
Keja, A.J.	Hoogovens - IJmuiden	NL
Kok, J.J.	T.H.D., Afd. WB, Mekelweg 2, Delft	NL
Knoll, R	Daimler Benz A.G., D7000 Stuttgart 60	FRG
Koga, H.	Honda Engineering Co. Sayama-Shi, Saitama 350-13	J

List of participants

NAME	ADDRESS	COUNTRY
Kramer, G.	Voltastraat 2, Terneuzen	NL
Landeweerd, J.A.	T.H.E., Afd. Bedrijfskunde Postbus 513, Eindhoven	NL
Levin, M.	Ind. Social Research, NTH Trondheim	N
Lübbert, U.	I.I.T.B., Seb. Kneippstr. 12 Karlsruhe	FRG
Maier, H.	Acc. der Wissenschaft DDR Leipzigerstr. 3, 108 Berlin	DDR
Margulies, F.	Deutschmeisterpl. 2 Vienna 1013	A
Mårtensson, L.K.	Lab. of Industr. Ergonomics KTH, Stockholm S-10044	S
Misul, M.	Olivetti SpA, Ivrea,	I
Müller, H.	Digitron A.G., Bruegg	CH
Muller, F.A.	United Nations, Palais des Nations 1211, Genève	CH
Mumford, E.	Manchester Business School Booth Street West, Manchester	UK
Myrvang, G.	Norwegian Computing Centre Forskningsvn 1B, Oslo 3	N
Nemes, L.	Hungarian Academy of Sciences	H
Nillensen, G.M.	P.O.Box 1800, Amsterdam	NL
Overakker, P.J.	Bedrijfskunde, L.H. Wageningen Oude Delft 23, Delft	NL
Paasche, T.	Inst. Psychol. + Soc. Research 7034 Trondheim - NTH	N
Pasmooij, C.K.	N.I.P.G., Postbus 124, Leiden	NL
Paternotte, P.H.	Afd. Bedrijfskunde, T.H.E. Postbus 513, Eindhoven	NL
Remmerswaal, J.L.	Metaalinst. TNO, Laan van Westenk 501, Apeldoorn	NL
Rijnsdorp, J.E.	Afd. CT, T.H.T., Postbus 217, Enschede	NL
Rookmaaker, D.P.	Nederlandse Spoorwegen, Afdl. Ergonomie, Utrecht	NL
Roveda, C.	Politecnico di Milano, Milano	I
Scapin, D.	IRIA, 172 Ave. de la République 92 Nanterre	F
Schneider, H.W.	Meet- en Regeltechniek T.H.D., Mekelweg 2, Delft	NL
Scholtens, S.	Hoogovens - IJmuiden	NL
Seewald, F.P.	KU - Nijmegen, Fazantenveld 149, Cuyck	NL

List of participants

NAME	ADDRESS	COUNTRY
Soede, M.	N.I.P.G., Postbus 124, Leiden	NL
Stassen, H.J.	Afd. WB, T.H.D., Mekelweg 2, Delft	NL
Sterk, V.	Hoogovens - IJmuiden	NL
Umbach	T.H.T., Afd. EL, Postbus 217, Enschede	NL
Verhagen, L.H.J.M.	T.H.T., Afd. CT, Postbus 217, Enschede	NL
De Vlaming, P.M.	N.I.P.G., Postbus 124, Leiden	NL
Wahlström, B.	Technical Res. Centr. Otakaari 5, 02150 Espoo 15	SF
Welfonder, E.	Universität Stuttgart, Pfaffenwaldring 23	FRG
Whiston, T.	Social Policy Research Univ. University of Sussex, Brighton	UK
Wieser, G.	Inst. Sociology, Univ. Vienna, Alserstrasse 33, 1080 Vienna	A
Wilson, A.T.M. Wilson, M.E.	ICQWL, 46 Clarence Terrace, London NW1 4RD	UK
Zeeb, H.	Hewlett Packard, G.m.b.H. Herrenbergerstr. 110, 7030 Boeblingen	FRG
Zonderman, R.	Hoogovens - IJmuiden	NL